Integrating Wireless Networks: Mesh, Ad Hoc and Radio Access Networks

Integrating Wireless Networks: Mesh, Ad Hoc and Radio Access Networks

Edited by **Frank Elliott**

CWILLFORD PRESS
New York

Published by Willford Press,
118-35 Queens Blvd., Suite 400,
Forest Hills, NY 11375, USA
www.willfordpress.com

Integrating Wireless Networks: Mesh, Ad Hoc and Radio Access Networks
Edited by Frank Elliott

International Standard Book Number: 978-1-68285-089-3 (Hardback)

Contents

Preface   **VII**

Chapter 1   **Congestion Avoidance in IP Based CDMA Radio
Access Network**   **1**
Syed Shakeel Hashmi

Chapter 2   **DQSB: A Reliable Broadcast Protocol Based on
Distributed Quasi-Synchronized Mechanism for
Low Duty-Cycled Wireless Sensor Networks**   **15**
Yun Wang, Peizhong Shi, Kai Li and Jie Wu

Chapter 3   **Channel Assignment Algorithms for
MRMC Wireless Mesh Networks**   **35**
Mohammad A Hoque and Xiaoyan Hong

Chapter 4   **Analysis of Wimax Physical Layer using Spatial Multiplexing**   **55**
Pavani Sanghoi and Lavish Kansal

Chapter 5   **Stochastic Analysis of Random Ad Hoc Networks with
Maximum Entropy Deployments**   **68**
Thomas Bourgeois and Shigeru Shimamoto

Chapter 6   **Impact of Macrocellular Network Densification on
the Capacity, Energy and Cost Efficiency in
Dense Urban Environment**   **88**
S. F. Yunas, T. Isotalo, J. Niemelä and M. Valkama

Chapter 7   **Group Session Key Exchange Multilayer Perceptron Based
Simulated Annealing Guided Automata and
Comparison Based Metamorphosed Encryption in
Wireless Communication (GSMLPSA)**   **108**
Arindam Sarkar and J. K. Mandal

Chapter 8   **Design of Star-Shaped Microstrip Patch Antenna for
Ultra Wideband (UWB) Applications**   **128**
Mustafa Abu Nasr, Mohamed K. Ouda and Samer O. Ouda

Chapter 9   **Data Collection Scheme for Wireless Sensor Network with Mobile Collector**   137
Khaled Almi'ani, Muder Almi'ani, Ali Al_ghonmein and Khaldun Al-Moghrabi

Chapter 10   **Virtual 2D Positioning System by using Wireless Sensors in Indoor Environment**   144
Hakan Koyuncu and Shuang Hua Yang

Chapter 11   **Optimization of Performance Metrics of Lar in Ad-Hoc Network**   159
Neelesh Gupta and Roopam Gupta

Chapter 12   **A Novel Architecture for SDN-Based Cellular Network**   175
Md. Humayun Kabir

Chapter 13   **An Efficient Model for Reducing Soft Blocking Probability in Wireless Cellular Networks**   190
Edem E. Williams and Daniel E. Asuquo

Permissions

List of Contributors

Preface

This book has been a concerted effort by a group of academicians, researchers and scientists, who have contributed their research works for the realization of the book. This book has materialized in the wake of emerging advancements and innovations in this field. Therefore, the need of the hour was to compile all the required researches and disseminate the knowledge to a broad spectrum of people comprising of students, researchers and specialists of the field.

Wireless networks are omnipresent in the modern age. The aim of this book is to contribute to the advancements in the field of wireless networks through comprehensive discussions on mesh networking, broadband access networks, routing, wireless multimedia systems, information access, etc. This book presents a practical approach ideal for professionals, but can also be used by students pursuing engineering, communication studies, etc. As this field is emerging at a fast pace, this book will help the readers to better understand the concepts of wireless networks.

At the end of the preface, I would like to thank the authors for their brilliant chapters and the publisher for guiding us all-through the making of the book till its final stage. Also, I would like to thank my family for providing the support and encouragement throughout my academic career and research projects.

Editor

Congestion Avoidance in IP Based CDMA Radio Access Network

Syed Shakeel Hashmi

Electronics and Communication Engineering,FST ICFAI University Dehradun,India
hashmi_deg@yahoo.com

Abstract

CDMA is an important air interface technologies for cellular wireless networks. As CDMA-based cellular networks mature, the current point-to-point links will evolve to an IP-based Radio Access Network (RAN). mechanisms must be designed to control the IP Radio Access Network congestion.

This Paper implements a congestion control mechanism using Router control and channel control method for IP-RAN on CDMA cellular network. The Router control mechanism uses the features of CDMA networks using active Queue Management technique to reduce delay and to minimize the correlated losses. The Random Early Detection Active Queue Management scheme (REDAQM) is to be realized for the router control for data transmission over the radio network using routers as the channel.

The channel control mechanism control the congestion by bifurcating the access channel into multiple layer namely RACH, BCCH and DCH for data accessing. The proposed paper work is realized using Matlab platform.

Key Words

TCP,RAN, congestion, channel control,MAC.

1. Introduction

The present wireless communication system is moving towards the IP enabled network, where the cellular services are integrated with IP network for the transmission of data. Such networks are generally termed as IP-RAN network. In this network the Transmission Control Protocol (TCP) is the most widely used method to achieve elastic sharing between end-to-end IP flows. At present the core network basically relies on end-system TCP to provide congestion control and sharing but this will not be acceptable in coming future because, to avoid time-out, each TCP connection requires few packets to be stored in the network, and most of that storage occurs in the router buffer which leads to congestion.

The consumers of the future will put new requirements and demands on services. However, the fact is that today we already use different kinds of multimedia services, i.e. services that to some extent combine pictures, motion and audio. These include TV, video and the Internet. Many of these applications have become fundamental elements of our lives because they fulfill basic needs, for example, communication with friends, escape from reality and last but not least simply enjoying ourselves.

As technology develops, we can satisfy these needs by using new tools, new applications and new personal devices. When utilizing these new personal tools and services to enrich our lives, while being mobile, we are using Mobile Multimedia applications. As new handsets, new technologies and new business models are introduced on the marketplace, new attractive multimedia services can and will be launched, fulfilling the demands. Because the number of multimedia services and even more so, the context in which the services are used is numerous, the following model is introduced in order to simplify and clarify how different services will evolve, enrich our lives and fulfill our desires.

Without sufficient storage in router, the time-out will give a poor performance to the end user and prevent sharing in network. Providing larger storage for large number of connections will cause too much latency. So, if latency is to be limited then the number of connections must be severely reduced.

With the increase in the data access using these protocol, demands for larger bandwidth in coming future. Increasing bandwidth may not be a suitable solution as it is economically non-advisable. The decrease in the resources may lead to congestion in the network resulting to complete collapsing of the network. A mechanism is hence required to overcome these problems so as to support larger data in the constraint resource to provide fair routing with least congestion.

2. Congestion

An important issue in a packet-switched network is **congestion.** Congestion in a network may occur if the load on the network i.e. the number of packets sent to the network is greater than the capacity of the network. Congestion may occur due to several reasons such as overloading the network, burst transmission, variable bit rate transmission etc. congestion reduces the performance of a network and to be controlled.

Congestion control refers to the mechanism or technique to keep the network load below the capacity limit. Congestion happens in any system due to the involvement of waiting, abnormality in the flow etc. In network congestion occurs because routers and switches have queues or buffers that hold the packets before and after processing.

For example, a router has an input queue and an output queue for each interface. When a packet arrives at the incoming interface, it undergoes three steps before departing,.

1. The packet is put at the end of the input queue while waiting to be checked.

2. The processing module of the router removes the packet from the input queue   once it reaches front of the queue and uses its routing table and the destination address to find the route.

3.   The packet is put in the appropriate output queue and waits its turn to be sent.

The two issues which result in congestion are;

1)   If the rate of packet arrival is higher than the packet processing rate, the input queues become longer and longer.

2)   If the packet departure rate is less than the packet processing rate, the output queues become longer and longer.

The problem of congestion is a considerable factor in the up coming IP-RAN network, where the data between two cellular terminals are communicated using wireless and router interface. The IP-RAN is proposed to be integrated with CDMA communication system so as to enable IP access in Wireless network.

3. Congestion Control Policies

In this implementation two congestion control mechanisms to maximize network capacity while maintaining good voice quality: *admission control, and router control mechanism are evaluated.* Call admission control in current CDMA cellular voice networks is restricted to controlling the usage of air interface resources.

The principle underlying both schemes is regulation of the IP RAN load by adjusting the admission control criterion at the air-interface. the impact of router control in the form of active queue management is also evaluated. IP routers using a drop tail mechanism during congestion could produce high delays and bursty losses resulting in poor voice quality. Use of active queue management at the routers reduces delays and loss correlation, thereby improving voice quality during congestion. Using simulations of a large mobile network,

3.1 Router Control Mechanism

3.1.1 Active Queue Management

A traditional Drop Tail queue management mechanism drops the packets that arrive when the buffer is full. However, this method has two drawbacks.

Firstly, this mechanism allows a few connections with prior request to dominant the queue space allowing the other flows to starve making the network flow slower. Second, Drop Tail allows queues to be full for a long period of time. During that period, incoming packets are dropped in bursts. This causes a severe reduction in throughput of the TCP flows. One solution to overcome is to employ active queue management (AQM) algorithms. The purpose of AQM is to react to incipient congestion before the buffer overflows. AQM allows responsive flows, such as TCP flows, to react timely and reduce their sending rates in order to prevent congestion and severe packet losses.

Active Queue Management (AQM) interacts with TCP congestion control mechanisms, and plays an important role in meeting today's increasing demand for quality of service (QoS). Random Early Detection (RED), is an enhanced Algorithm employing Active Queue Management (AQM) scheme for it's realization. It is a gateway-based congestion control mechanism. An accurate model of TCP with RED can aid in the understanding and prediction of the dynamical behavior of the network. In addition, the model may help in analyzing the system's stability margins, and providing the design guidelines for selecting network parameters. These design guidelines are important for network designers whose aim is to improve network robustness. Therefore, modeling TCP with RED is an important step towards improving the network efficiency and the service provided to Internet users.

3.1.2 Random Early Detection Router

Random early Detection is one of the active queue management control mechanism deployed at gateways[2]. The RED gateway detects incipient congestion by computing the average queue size (Jacobson, 1998). The gateway could notify connections of congestions either by dropping packets arriving at the gateway or by setting a bit in packet headers. When the average queue size exceeds a preset threshold, the gateway drops or marks each arriving packet with a certain probability, where the exact probability is a function of the average queue size. RED gateways keep the average queue size low while allowing occasional burst of packets in the queue. Figure 1 show a network that uses RED gateway with a number of source and destination host. The RED congestion control mechanism monitors the average queue size for each output queue, and using randomization, chooses connections to notify of the congestion Transient congestion is accommodated by a temporary increase in the queue. Longer-lived congestion is reflected by an increase in the computed average queue size and result In randomized feedback to some of the connections to decrease their windows. The probability that a connection is notified of congestion is proportional to that connection's share of the throughput through the gateway.

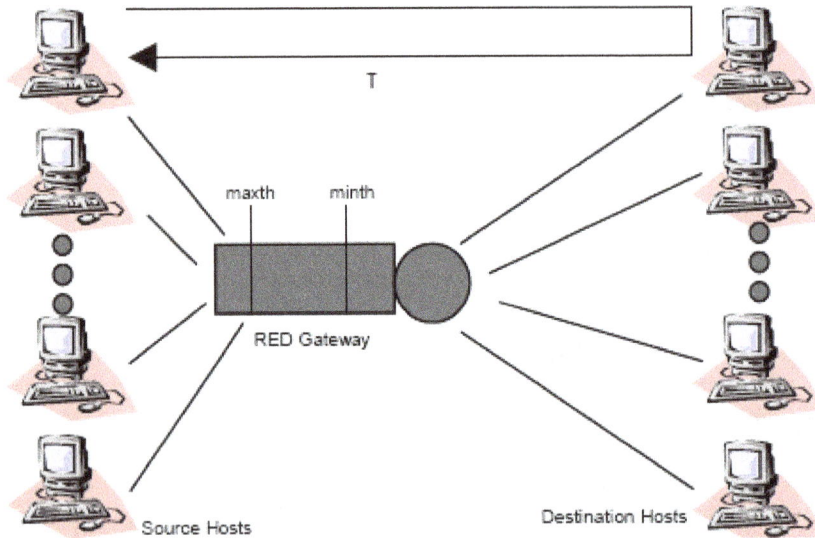

Fig:1 A network with RED gateway

3.1.3 Red Algorithm

RED mechanism contains two key algorithms. One is used to calculate the exponentially weighted moving average of the queue size, so as to determine the burstiness that is allowed in the gateway queue and to detect possible congestion. The second algorithm is for computing the drop or marking probability, which determines how frequently the gateway drops or marks arrival packets. This algorithm can avoid global synchronization by dropping or marking packets at fairly evenly spaced intervals. Furthermore, sufficiently dropping or marking packets, this algorithm can maintain a reasonable bound of the average delay, if the average queue length is under control.

The Random Early Detection (RED) algorithm is given as[6],

Let '*wq*' be the weight factor and *qk+1* be the new instantaneous queue size.

'*q̄*' is the average queue size and '*q*' is the instantaneous queue size. Then at every packet arrival, the RED gateway updates the average queue size as

$$\overline{q}_{k+1} = (1 - w_q) \cdot \overline{q}_k + w_q \cdot q_{k+1} \quad \text{-------- (3.1)}$$

During the period when the RED gateway queue is empty, the system will estimate the number of packets '*m*' that might have been transmitted by the router. So, the average queue size is updated as

$$\overline{q}_{k+1} = (1 - w_q)^m \cdot \overline{q}_k, \quad \text{--------- (3.2)}$$

where

m=idle_time / transmission_ time

Where *idle_time* is the period that the queue is empty and *transmission_time* is the typical time that a packet takes to be transmitted.

The average queue size is compared to two parameters: the minimum queue threshold *qmin*, and the maximum queue threshold *qmax*. If the average queue size is smaller than *qmin*, the packet is admitted to the queue. If it exceeds *qmax*, the packet is marked or dropped.

If the average queue size is between *qmin* and *qmax*, the packet is dropped with a drop probability *pb* that is a function of the average queue size.

$$p_{b_{k+1}} = \begin{cases} 0 & if \quad \overline{q}_{k+1} \leq q_{min} \\ 1 & if \quad \overline{q}_{k+1} \geq q_{max}, \\ \dfrac{\overline{q}_{k+1} - q_{min}}{q_{max} - q_{min}} \cdot p_{max} & otherwise \end{cases}$$

-------------- (3.3)

where *pmax* is the maximum packet drop probability.

The final drop probability '*pa*' is given by

$$p_a = \frac{p_b}{1 - count \cdot p_b}$$

------------- (3.4)

C*ount* is the cumulative number of the packets that are not marked or dropped since the last marked or dropped packet. It is increased by one if the incoming packet is not marked or dropped. Therefore, as *count* increases, the drop probability increases. However, if the incoming packet is marked or dropped, *count* is reset to 0.

3.1.4 Red Method

For each packet arrival

    calculate the average queue size *avg*

    if *minth <= avg < maxth*

    calculate the probability *pa*

    with probability *pa*:

    mark the arriving packet

else if *maxth <= avg*

mark the arriving packet.

3.1.5 Red Drop Probability (P$_a$)

pb = maxp x *(avg - minth)/(maxth - minth)* [3.5]

    where

    pa = pb/ (1 - count x *pb)* [3.6]

pb = pb x PacketSize/MaxPacketSize [3.5a]

3.1.6 Avg -Average Queue Length

$avg = (1 - wq) \times avg + wq \times q$

Where q is the newly measured queue length.

This *exponential weighted moving average* is designed such that short-term increases in queue size from bursty traffic or transient congestion do not significantly increase average queue size.

3.2 channel control method

A medium access control (MAC) protocol is developed for wireless multimedia networks based on frequency division duplex (FDD) wideband code division multiple access (CDMA).

In order to admit real-time connections, a Connection Admission Control (CAC) scheme is needed. One of the challenging issues of a CAC algorithm is that the bandwidth requirement of a connection cannot be determined at connection admission time due to the traffic burst. Thus, effective bandwidth-based CAC schemes are generally used to solve this problem.

3.2.1 Channel Admission Control (CAC)

The congestion occurring in CDMA can be controlled by proper controlling of channel accessing using the channel control methods[12],[13]. For the controlling of channel accessing in cdma 2000 a MAC protocol is proposed. In the MAC protocol following are the transport channels used:

1) Random access channel (RACH) is used by mobile terminals to send control packets,

2) Broadcast control channel (BCCH) conveys system information from the base station to mobile terminals, and

3) Dedicated channel(DCH) is a point-to-point channel used to transmit data from mobile terminals to the base station or vice versa. The different transport channels are multiplexed in the code division.

Many transport channels are defined for wide-band FDD-CDMA networks in 3GPP specifications. The packet transmission in dedicated channels (DCHs) is focused in this implementation. Since signaling channels are required to assist DCHs, I also consider two other transport channels, i.e., the random access channel (RACH) to send requests from a mobile terminal to the base station and the broadcast channel (BCH) to send feedback of resource allocation from the base station to mobile terminals. Mappings of DCH, BCH, or RACH onto a physical channel.

4. Design Approach

The implementation is made on the communication system following cdma architecture[5]. The transmitter and the receiver section is linked via wireless access using wireless medium and the router unit. Figure 2 shows the proposed link model for CDMA IP RAN access network.

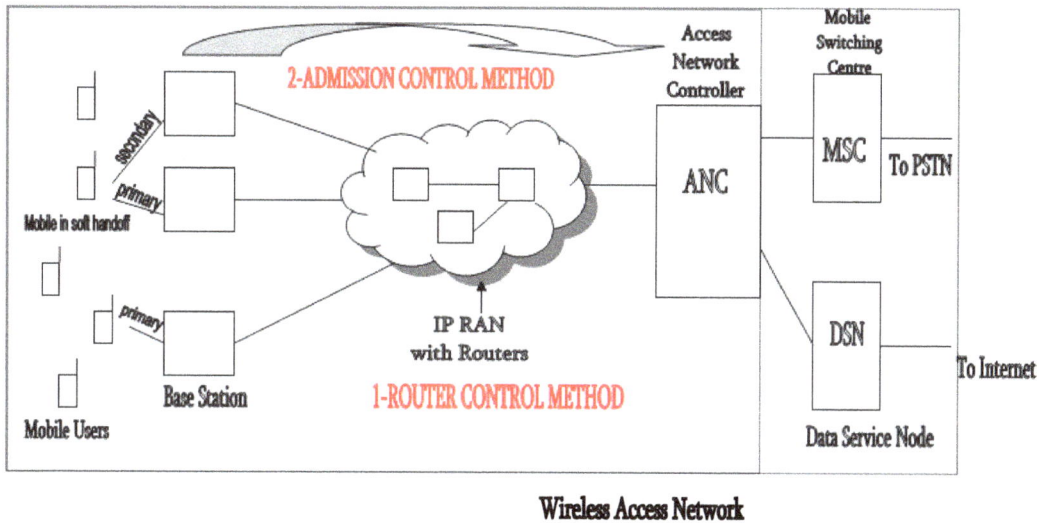

Wireless Access Network
Fig:2 CDMA Wireless Access Network with IP RAN

4.1 MAC Controller

The MAC controller module controls the channel access by dividing the channel into three sub-channelas;

• Random access channel *(RACH)*. This channel is used by mobile terminals to send control packets, e.g., connection requests.

• Broadcast control channel *(BCCH)*. This channel is a point-to-multipoint channel, which is used to convey system information from the base station to mobile terminals. For example, the feedback about resource allocation is transmitted in this channel.

• Dedicated channel *(DCH)*. The DCH is point-to-point channel, which is used to transmit data from mobile terminals to the base station or vice versa. DCHs, RACH and BCCH are multiplexed in the code division. In DCHs of FDD mode wideband CDMA, both MC transmission and VSF transmission are used. A DCH can have variable transmission rate depending on the spreading factor, and the basic transmission rate of the DCH corresponds to the maximum spreading factor used in this channel. Thus, a variable-length MAC packet is accommodated in a DCH.

4.1.1 MAC Transmitting Operation

When a mobile terminal wants to communicate, it needs to send a connection request through RACH to the receiver terminal. Once this request is received at the receiver base station, based on the availability of the receiver an acknowledgment is generated through BCCH sending a high signal to the transmitter unit. On reception of acknowledgment the transmitter generates a data transmission request, and on confirmation the data is transmitted through DCH channel.

Connection in RACH is requested

Feedback BCCH='1'

Request for connection

Connection is rejected

Connection is admitted

Packet transmission is over

No Resonance

Request for transmission

Feedback BCCH='1'

Packet transmission

Packet in DCH

Transmission Request in RACH

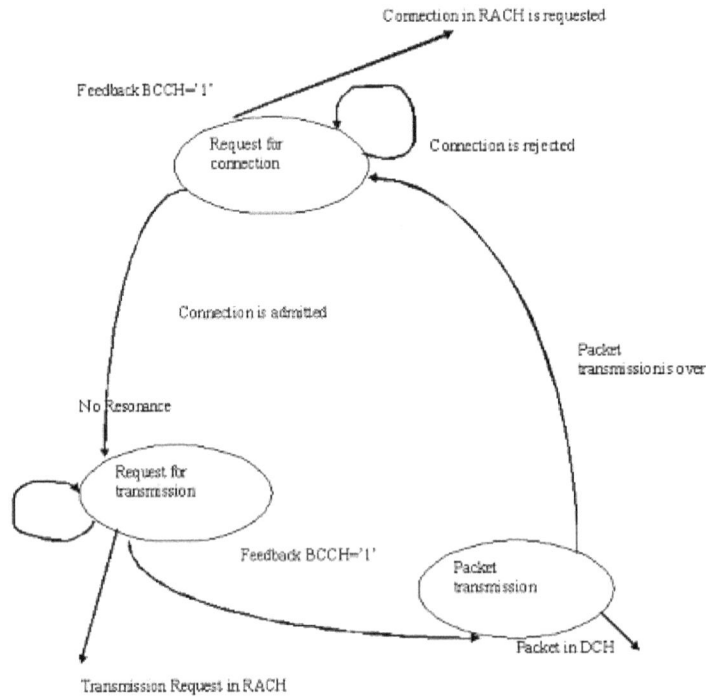

Fig:3 The operation of the mac controller is illustrated as state diagram

4.1.2 MAC Receiving Operation

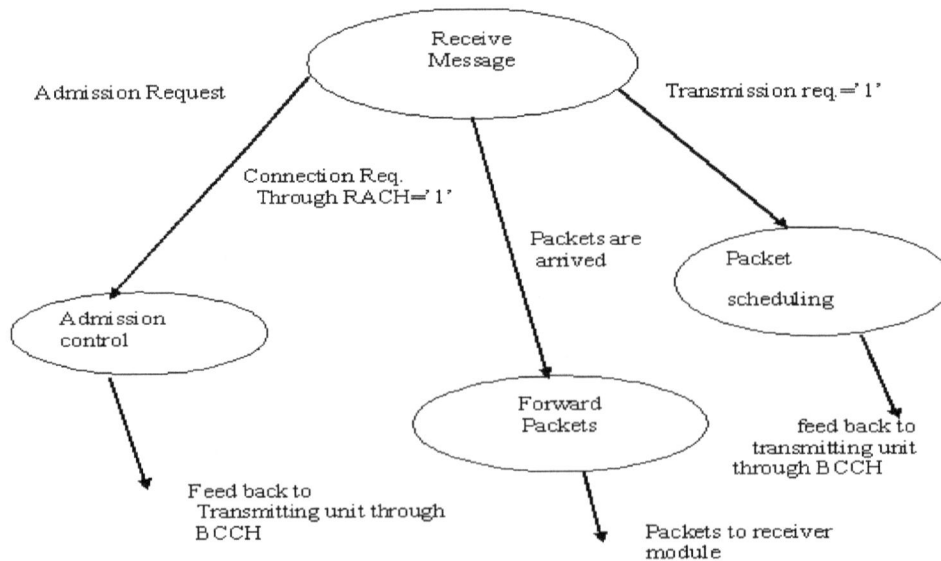

Receive Message

Admission Request

Transmission req.='1'

Connection Req. Through RACH='1'

Packets are arrived

Packet scheduling

Admission control

Forward Packets

feed back to transmitting unit through BCCH

Feed back to Transmitting unit through BCCH

Packets to receiver module

Fig:4 MAC Receiving Operations

The receiver controls the accessing using an state diagram as shown in figure 4. The receiver respond to the request signal generated from the transmitting section and accept the data from transmitter forwarding to the receiver unit using he CAC control mechanism.

4.2 Router Module

This section gives the architecture designed to implement the buffer management algorithm of Random Early Detection (RED). The architecture consists of four main components: Main Controller, Packet Drop Probability Unit (PDPU), Random Packet Drop Unit (RPDU) and the Compute Random Value Unit (CRVU). Figure shows a system architecture overview

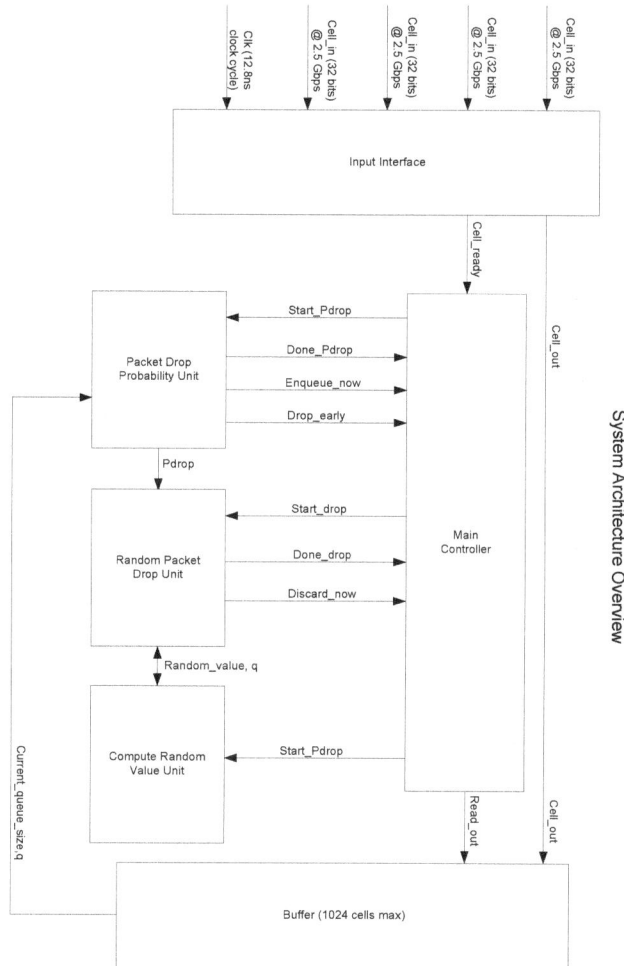

Fig:5 RED System Architecture Overview

4.2.1 Input Interface

RED is essentially a packet-discard algorithm. There are four concurrent connections to the input interface as shown in figure 5. One cell arrives at an input connection every four clock cycles i.e. a total of four cells arrive at the interface. Sixteen clock cycles constitute a time slot. Four cells arrive at one input connection every time slot. Thus 16 cells arrive at the input interface every time slot.

4.2.2 Main Controller

The Main Controller controls the entire system. It issues signals that control the Packet Drop Probability Unit, Compute Random Value Unit and Random Packet Drop Unit. The Main Controller is activated only when a cell arrives at the input interface and a *cell_ready* signal is issued by it.

In addition, the Main Controller is responsible for interacting with the Buffer. It updates the current queue size every time slot. Two cells leave the buffer every time slot and the Main Controller reduces the current queue size by two every sixteen clock cycles. Moreover when a packet has been determined to be enqueued the current queue size is incremented.

4.2.3 Packet Drop Probability Unit

This module calculates the probability from the obtained q- size and decides whether to accept or drop the packets received. PDPU consists of three look up tables. The first look up table contains computed values of the average queue size, $avg \leftarrow (1 - w)^m \, avg$ for corresponding packet transmission time, m. The second contains the initial packet drop probability, $p_b \leftarrow max_p \, (avg - min_{th}) / (max_{th} - min_{th})$ for corresponding values of avg. And the third look up table includes final packet drop probability,

$p_a \leftarrow p_b / (1 - count \, p_b)$, for corresponding values of p_b.

4.2.4 Configure Random Value Unit

The Configure Random Value Unit uses a random generator function to compute a random number. An 8 bit initial seed is used and set equal to "00000010". Tests were conducted with different seeds and a seed of "00000010" produced the most unbiased random number i.e. random numbers were never repeated, a unique one was output each time..

A feedback signal is computed by running an XOR process on each of 8 bits of the initial seed in the following manner:

4.2.5 Random Packet Drop Unit

 One point to note here is the fact that only packets that were determined by the PDPU to have an average queue size between the minimum and maximum thresholds are marked. Other packets do not interact with this unit they are chiefly concerned with the PDPU.

Analysis Of Complete System

For the analysis of the complete system we need the below flowchart

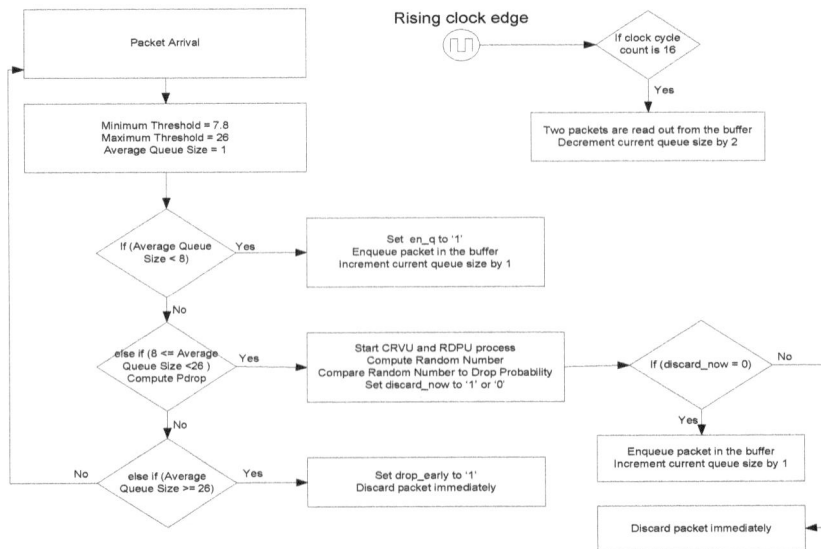

Figure 6 Summary of all processes in the RED implementation

Figure6   shows that whenever count reaches 16 i.e. a time slot is over, two packets from the buffer are read out and thus the current queue size of the buffer is reduced by two. In the first case when the clock count reaches 16 the current queue size decreases by 2.

5. Results And Analysis

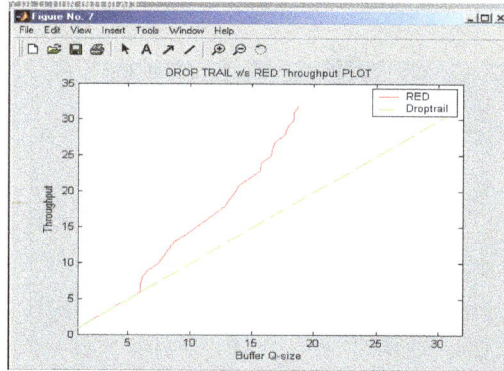

Fig: 7 Throughput v/s Q-size for Droptail and RED method

Fig illustrates the throughput performance of the Router control mechanism for RED and Droptail method. It can be observe that the throughput remain same for RED & DT until minimum threshold.After min threshold the throughput of RED router is seem to be improved because of packet dropping .

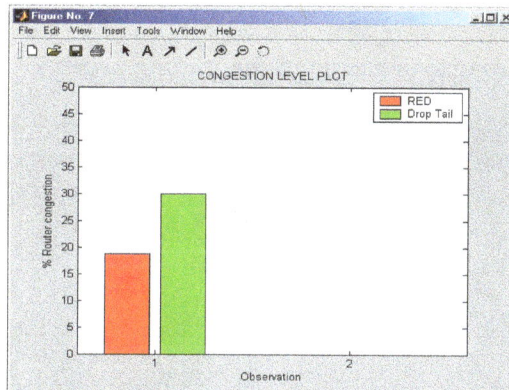

Fig:8 The congestion level plot for Droptail and RED method

Fig shows the congestion level obtained for the simulation of implemented design.It can be seen that RED gives congestion avoidance of about 40% compare to the Droptail method.

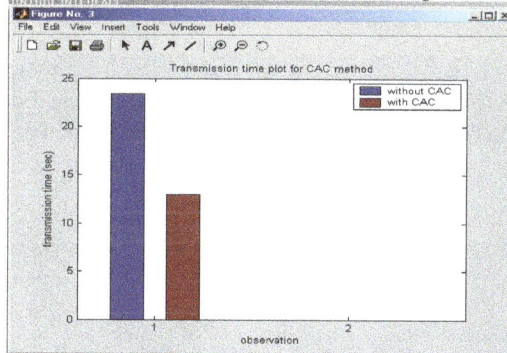

Fig: 9 Transmission time comparisons for CDMA base comm system with and without CAC

Fig shows the transmission time comparison plot for two methods used for controlling the congestion in Air interface.From the plot it can be observed that about 50% of transmission time could be saved using CAC method than the existing method.

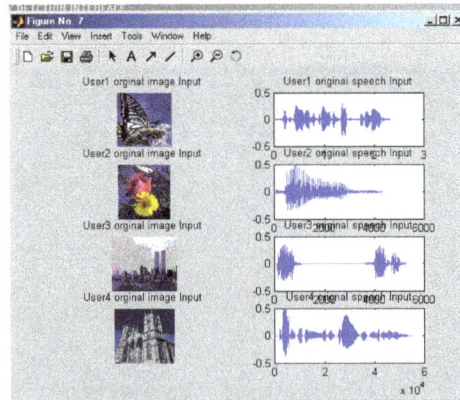

Fig:10 The image and speech samples considered for processing and evaluating the implemented system.

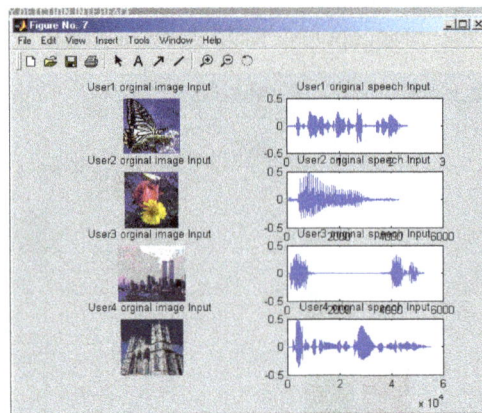

Fig11 Retrived outputs of image and speech sequence for four users

CONCLUSION

The evolution of wireless communication for multi bit rate application has come down to integration of different technologies for efficient transmission of data over wireless network.

The IP-RAN network is one such result of this technology integration. The evolved IP network though proposed to support multi bit rate application can fails in providing efficient performance for bursty network. The network fails to provide better performance due to limitations and resource such as bandwidth resulting in congestion.

This paper realizes the two most advanced congestion control policies namely the Channel Admission Control and Router Control controlling in wireless network. For the evaluation of control policies a wireless network with CDMA2000 standard is adapted.

The system realizes follows the CDMA2000 standard with CDMA transceiver architecture. The transceiver realizes are integrated with the Router and Channel control mechanism to overcome the congestion happening at the network. The CDMA architecture realized for communication is modular approached with modules such as scramble code generator using LFSR logic spreader spreading the user data over the entire bandwidth for the BW spreading of user data a OVSF spread code is realized using Walsh Hadamard coding. The multiple spreaded user data are then bundle over power spectrum modulated using BPSK modulation before transmission .A multi-user detection logic is realized at the receiver end using matched filter concept to the isolation of user data before dispreading to retrieve back the final data.

A router interface is realized as an intermediate interface between the source and destination to provide better services using an IP- enabled wireless. The congestion control policy for router control mechanism is realized using the implementation of RED algorithm. This algorithm is designed an incorporated with router architecture for the control of congestion at early stage. The controlled is achieved by floating the congestion level between minimum and maximum threshold.

An air interface control mechanism is also realized for the control of congestion occurring at the wireless channel. To avoid the congestion at channel the channel access mechanism is developed. In this mechanism the channel is bifurcated into dedicated layers using frequency division duplex, where each layer of a channel is dedicated for a particular operation named as RACH, DCH and BCCH.For the evaluation of two algorithms, the algorithms are compared with their existing control methods.

From the implementation and observation it is seen that the performance can be improved for IP-RAN with the integration Router and Channel control mechanism from its counterparts.

The Router control mechanism can be embedded with the transceiver architecture provides 30 to 40% reduction in congestion level at the router making the transmission of the image and speech data faster. It is observed that Router design with RED algorithm provides better throughput than the Droptail method. The Channel Admission Control method realized shows a performance enhancement of about 50% faster than the earlier system.

From all the observation made, it could be concluded that CDMA system with Router control mechanism using RED algorithm and Channel control mechanism using MAC protocol can enhance the performance of the data transmission for both image and speech in wideband CDMA network.

FUTURE SCOPE

The implementation make few assumptions for the implementation of the two control mechanisms such as idle channel equal speed traffic balanced, burstiness etc.These assumptions hold well during algorithm evaluation but may deviate in its operation in real time scenario. An effort could be made to evaluate the performance considering these factors also.

The proposed implementation is evaluated on apart of image and speech sample, an effort can be made to speed up the operation of this implementation in future.

The algorithm can also be tried out with other communication system such as MC-CDMA, CS-CDMA etc.

The implementation can also be evaluated considering various channel parameters for real time evaluation.

REFERNCES

1. Andreas F. Molisch,"Wideband Wireless Digital Communications", Pearson Education, 2001, PP-18.

2. A. A. Akintola, G. A. Aderounmu,and L. A. Akanbi Obafemi Awolowo University, Ile-Ife, Nigeria, M. O. Adigun University of Zululand,Kwadlangezwa, Republic of South Africa," Modeling and

Performance Analysis of Dynamic Random Early Detection (DRED) Gateway for Congestion Avoidance", Issues in Informing Science and Information Technology,PP-623-633.

3. Behrouz A. Fourouzan,"Data Communication and Networking", third edition, Tata McGRAW-HILL, 2004.

4. Erik Dahlman, Per Beming, Jens Knutsson, Fredrik Ovesj¨o, Magnus Persson, and Christiaan Roobol," WCDMA—The Radio Interface for Future Mobile Multimedia Communications", IEEE TRANSACTIONS ON VEHICULAR TECHNOLOGY, VOL. 47, NO. 4, NOVEMBER 1998, PP-1105, 1115.

5. G. Heijenk, G. Karagiannis, V. Rexhepi, and L. Westberg, "Diffserve resource management in ip-based radio access networks," in Proceedings of WPMC, Aalborg, Denmark, September 2001.

6. Sneha Kumar Kasera, Ramchandran R, Sandra and Xin Wang," Congestion Control Policies for IP-Based CDMA Radio Access Network,"IEEE Trans on mobile computing, vol4, pp.349-62, july/aug2005.

7. S.Floyd "Connections with multiple congested gateways in packet switched network. Part I.One way traffic". Comput. Commun. Rev. vol.21 no.5 pp 30-47. Oct.1991

8. S.Floyd and V.Jacobson,"Random Early Detection Gateways for Congestion Avoidance", IEEE/ACM Trans.Networking, vol.1, no.4, pp.397-413, Aug1993.

9. Theodre S. Rappaport, Joseph C. Liberty," Smart Antennas for Wireless Communications IS-95 and third generation CDMA Applications", Prentice Hall PTR.

10. Vijay K. Garg,"Wireless Network Evolution", Pearson Education, 2003.

11. Vijay K. Garg,"IS-95 CDMA and cdma2000", Pearson Education, 2002.

12. Xudong Wang, *Member, IEEE,*" Wide-Band TD-CDMA MAC With Minimum- Power Allocation and Rate- and BER-Scheduling for Wireless Multimedia Networks" IEEE/ACM TRANSACTIONS ON NETWORKING, VOL. 12, NO. 1, FEBRUARY 2004, PP.103,105.

13. Xudong Wang, Member, IEEE," An FDD Wideband CDMA MAC Protocol with Minimum-Power Allocation and GPS-Scheduling for Wireless Wide Area MultimediaNetworks",IEEETRANSACTIONSONMOBILECOMPUTING, VOL. 4, NO. 1, JANUARY/FEBRUARY 2005,PP-16,24

14. "www.Ericsson.com" Ericsson Radio Systems AB," Basic Concepts of WCDMA Radio Access Network", white paper.

DQSB: A Reliable Broadcast Protocol Based on Distributed Quasi-Synchronized Mechanism for Low Duty-Cycled Wireless Sensor Networks

Yun Wang[1] and Peizhong Shi[1] and Kai Li[1] and Jie Wu[2]

[1]School of Computer Science and Engineering, CNII, Southeast University, Nanjing, 211189, P.R. China
{yunwang, peizhongshi, newlikai}@seu.edu.cn
[2]Department of Computer and Information Sciences, Temple University, Philadelphia, PA 19122
jiewu@temple.edu

ABSTRACT

In duty-cycled wireless sensor networks, deployed sensor nodes are usually put to sleep for energy efficiency according to sleep scheduling approaches. Any sleep scheduling scheme with its supporting protocols ensures that data can always be routed from source to sink. In this paper, we investigate a problem of multi-hop broadcast and routing in random sleep scheduling scheme, and propose a novel protocol, called DQSB, by quasi-synchronization mechanism to achieve reliable broadcast and less latency routing. DQSB neither assumes time synchronization which requires all neighboring nodes wake up at the same time, nor assumes duty-cycled awareness which makes it difficult to use in asynchronous WSNs. Furthermore, the benefit of quasi-synchronized mechanism for broadcast from sink to other nodes is the less latency routing paths for reverse data collection to sink because of no or less sleep waiting time. Simulation results show that DQSB outperforms the existing protocols in broadcast times performance and keeps relative tolerant broadcast latency performance, even in the case of unreliable links. The proposed DQSB protocol, in this paper, can be recognized as a tradeoff between broadcast times and broadcast latency. We also explore the impact of parameters in the assumption and the approach to get proper values for supporting DQSB.

KEYWORDS

Wireless Sensor Networks, Duty-cycled, Broadcast, Routing, Asynchronous Sleep Scheduling, Quasi-Synchronization

1. INTRODUCTION

A Wireless Sensor Network (WSN) consists of a large number of small and low cost sensor nodes powered by small batteries and equipped with various sensing devices to observe events in the real world [1-4]. Usually, for many applications, once a WSN is deployed, probably in an inhospitable terrain, it is expected to gather required data for a certain period of time, which can reach a length of years. To bridge the gap between limited energy supplies and network lifetime, a WSN has to operate in a low duty-cycled manner, where nodes schedule themselves to be active for a brief period of time and then stay asleep for a long period of time [5, 6]. There are two types of duty-cycled WSNs, i.e. asynchronous sleep scheduling, where each sensor keeps a sleep schedule independent of another, and synchronous sleep scheduling, where sensors make synchronized periodic duty cycling with their neighboring nodes to support broadcast or unicast and reduce the idle listening energy cost. Any sleep scheduling scheme has to ensure that data can always be routed from source to sink [7].

Usually, sleep schedules are completely uncoordinated. Due to the variation of awake time and duration of the active interval, the whole network is more than often disconnected, and delay encountered in packet delivery due to loss in connectivity can become a critical problem. As a result, a path from source to sink may not always be available, and a sufficient number of nodes have to remain awake to ensure the existence of such a path. Consequently, data is stored at a node till its proper neighboring node wakes up and delivers the data to the sink. This approach would delay the delivery of messages to a sink considerably.

The existing works based on synchronization assume that there are usually multiple neighbors available at the same time to receive the multicast/flooding message sent by a sender. This is not true in low duty-cycled asynchronous WSNs. Furthermore, synchronization is another issue that is difficult to achieve, especially over multiple hops. Periodic synchronization messages may become costly. Usually, synchronization protocols are complex and difficult to implement in large scale WSNs. Without synchronized sleep scheduling, B-MAC [8], WiseMAC [9] and X-MAC [10] are based on asynchronous sleep intervals and proven to be energy-efficient in scenarios with low or varying traffic loads. Unfortunately, they cannot be directly applied to broadcast applications because of their design intentions for unicast.

Multi-hop broadcast is an important network service in WSNs, especially for applications such as code update, remote network configuration, route discovery, and so on. Distinguished from the broadcast problem in always-on networks, two additional features make multi-hop broadcast in low duty-cycled WSNs become a new challenging issue. Firstly, a node which broadcasts a message once cannot guarantee that the message is received by all of its neighboring nodes simultaneously, while this property is satisfied in an always-on network. To successfully broadcast a message, a sender has to transmit the same message more than once if other nodes do not wake up at the same time. Essentially, broadcasting in such a network is implemented by a number of unicasts. Secondly, in asynchronous duty-cycled WSNs, each node cannot be aware of its neighboring nodes' sleep schedules without neighboring discovery and information exchange protocols which require nodes to remain awake for enough time in order to aware their neighbors' sleep schedules.

Therefore, a question arises: Is it possible to maintain a high broadcast delivery rate and to exploit nodes' sleep schedules without the support of synchronization protocol at the same time, in asynchronous duty-cycled WSNs, where each sensor turns on and off independently and network connectivity is intermittent? Different from the existing related work, we propose a quasi-synchronization mechanism in order to coordinate nodes' duty-cycled behaviors in a distributed manner. It is quasi because nodes are not required to wake up at the exact same time. Sleep schedule adjustments stop if all the nodes except the sender are able to receive broadcast messages.

The main contributions of this paper are summarized as follows: (1) We propose a novel protocol DQSB by a mechanism of quasi-synchronization for multi-hop broadcast, which neither assumes time synchronization that requires all neighboring nodes wake up at the same time, nor assumes duty-cycled awareness that is difficult to use in asynchronous WSNs; (2) After broadcast process from a sink is finished under the quasi-synchronization mechanism, other nodes can build their paths to the sink for transmitting their sensed data after receiving the broadcast messages. Moreover, these paths exhibit less latency because of no or very little waiting time; (3) We develop a simulator based on the ONE simulator [11] and evaluate DQSB, including broadcast times and latency in different duty cycles, the impact of network size, reliability with unreliable links and less latency routing paths for reverse data collection from each node to broadcast source node, such as a sink. Simulation results show that the performance of DQSB satisfies the design goals.

The rest of the paper is organized as follows: Section 2 reviews the related work. Section 3 describes the models and assumptions of the solution to broadcasting and routing in duty-cycled WSNs. The design and implementation of DQSB are presented in Section 4. Simulation results are discussed and analyzed in Section 5, where the impact of parameters in the assumption and the approach to get proper values are explored for supporting DQSB. We conclude the paper in Section 6.

2. RELATED WORK

As we addressed in Section 1, multi-hop broadcast plays an important role in WSNs. Compared with the problem of broadcasting in always-on networks, neighbor connectivity becomes a more difficult problem in duty-cycled WSNs, where each node stays awake only for a fraction of time and neighboring nodes are not simultaneously awake for receiving data. A bunch of literature has addressed this problem.

According to the mechanism supported broadcasting, existing solutions are put into two categories, including synchronous and asynchronous sleep schedules. The former, such as S-MAC [12] and T-MAC [13], simplifies broadcast communication by letting neighboring nodes stay awake simultaneously. The latter solution has become increasingly attractive for data communication because of its energy efficiency. Due to space limitation, we focus only on reviews for broadcast solutions for asynchronous duty-cycled WSNs.

The protocols B-MAC [8], WiseMAC [9], and X-MAC [10] are based on asynchronous wake-up intervals and have proven to be more energy-efficient in scenarios with low or varying traffic load. B-MAC supports single-hop broadcasting in the same way for unicast, since the preamble transmission over an entire sleep period gives all of the transmitting nodes' neighbors a chance to detect the preamble and remain awake for data packets. B-MAC and WiseMAC broadcasting are each energetically costly and inefficient. When transmitting a frame, a full preamble is appended for alerting neighboring nodes to stay awake for the upcoming transmission of the broadcast frame. This broadcast approach with a full preamble wastes a lot of energy for sending and receiving, while the actual data transmission is often comparatively short. Without control measures for forwarder selection in multi-hop flooding, every broadcast message to be rebroadcast by every node will experience the wireless-channel characteristic broadcast storm problem. Consequently, the broadcast success ratio and latency performance decreases. X-MAC, a low power MAC protocol, substantially improves B-MAC's excess latency at each hop and reduces energy usage at both the transmitter and receiver by employing a shortened preamble approach. But broadcast support is not clearly discussed in that paper. X-MAC is not promising for broadcasting since the transmitter has to continually trigger the neighbors to wake up, no matter whether it has received or not.

The (k)-Best-Instants broadcast algorithm [14], calculating the best instants and transmitting the frame with a minimized preamble, can be more efficient than using a costly full-cycle preamble like WiseMAC. Its assumption is the sender is aware of their neighbors' individual schedules. Wang et al. [15] transformed the problem into a shortest-path problem with the same assumption of duty-cycle awareness, which makes it difficult to use it in asynchronous WSNs since duty-cycle awareness needs periodic time-synchronization due to clock drifting. Focusing on energy-harvesting networks, Gu et al. [16] introduce the proactive generic delay maintenance algorithm to minimize the amount of energy while satisfying an end-to-end delay bound specified by application requirements for sink-to-many communications in energy-harvesting networks. But nodes in the network must share their duty-cycled working schedules with neighboring nodes for the assumption of duty-cycle awareness, so as to know when they can send a packet to their neighbors with the support of local synchronization techniques [17].

Opportunistic routing and data forwarding in low duty-cycled networks have acquired a lot of attention in recent years [18, 19]. But none of these solutions investigates the broadcasting. ADB [20] and opportunistic flooding [21] were designed with a gossiping approach as long as the network is connected. ADB avoids the problems with B-MAC and X-MAC by dynamically optimizing the broadcasting at the level of transmission, to each individual neighboring node. It allows a node to go to sleep immediately when no more neighbors need to be reached and does not occupy the medium for a long time, in order to minimize latency before forwarding a broadcast. The effort in delivering a broadcast packet to a neighbor is adjusted based on link quality, rather than transmitting throughout a duty cycle or waiting through a duty cycle for neighbors to wake up. Basically, ADB belongs to the unicast replacement approach and it needs significant modification to existing MAC protocols for supporting broadcast. In [21], a design of opportunistic flooding has been proposed for low duty-cycled networks with unreliable wireless links and predetermined working schedules. It provides probabilistic forwarding decisions at a sender based on the delay distribution of next-hop nodes and a forwarder selection method to alleviate the hidden terminal problem and a link-quality-based backoff method to resolve. However, these protocols for duty-cycled WSNs, belonging to the unicast replacement approach for supporting broadcasting, mostly focus on unicast communication and cannot well support broadcasting since one-hop broadcasting in such cases means to deliver data multiple times to all neighbors, which may lack efficiency in large scale networks, and also lack energy efficiency in delivering large chunks of data for broadcasting.

Hybrid-cast [22, 23], with low latency and reduced message count, overcomes the disadvantages of replacement via pure unicast. Under Hybrid-cast, nodes must switch their wake-up schedule to stay awake for enough time slots for neighbor discovery and information exchange. Then, the online forwarder selection algorithm works and helps to reduce the broadcast count or redundant transmission for multi-hop broadcast.

In conclusion, the above protocols either prevent themselves from being widely used in realistic environments due to their assumptions, including duty-cycled awareness and neighbor discovery supporting in asynchronous duty-cycled WSNs, or only belong to the unicast replacement approach for supporting broadcasting. We focus on exploiting nodes' sleep schedules and make adjustment strategies to solve the multi-hop broadcast problem by a distributed and quasi-synchronized manner. Meanwhile, a broadcasting node, such as sink, is in charge of data collection broadcast periodically. Receiving the broadcast messages, other nodes can build their paths to the sink for transmitting their sensed data, where these paths have less latency since the advantage of our quasi-synchronization mechanism. Unlike B-MAC, WiseMAC, and X-MAC, a unicast message in our paper can be transmitted along a path learned from the quasi-synchronized broadcasting and eliminates the waiting time for both transmitter and receiver.

3. PROBLEM DESCRIPTION

3.1. System Model

Suppose that in duty-cycled WSNs, there are p sensor nodes $N = \{n_1, n_2, \cdots\cdots, n_p\}$, $|N| = p$, working in two states: active and sleep states. All nodes have their own sleep schedules and are able to adjust their sleep schedules if necessary. A network is denoted by a time-dependent graph $G(t) = (N, E(t))$, where N is a complete set of nodes in the network and $E(t)$ is a set of undirected edges at time t. Evolving graph [24] is used to capture the dynamic characteristics, especially node intermittent connectivity with its neighbors. An evolving graph $G(t) = N(N, E(t))$ is connected during t, where $t \in [0, T]$ and T is one cycle length, if no isolated edge and vertex exists in $G(t)$. Let $L_{i,j}(t_{i,j}, T_{i,j})$ denote the intermittent connective link between nodes n_i and n_j, n_i, $n_j \in N$ and $L_{i,j}(t_{i,j}, T_{i,j}) \in E(t)$. The link begins at $t_{i,j}$ and keeps a period of time $T_{i,j}$. For its bidirectional property, $L_{i,j}(t_{i,j}, T_{i,j}) = L_{j,i}(t_{j,i}, T_{j,i})$ is established and $L_{i,j}(t_{i,j}, T_{i,j}) \in E(t)$. If n_i wakes up earlier than n_j, i.e. $t_i^a < t_j^a$, then $t_{i,j} = t_j^a$. Otherwise, $t_{i,j} = t_i^a$. All the variables used in the paper are listed in Table 1.

Table 1. Variables and their significance.

Variable	Description
t_i^a	Node n_i's time to wake up
t_i^s	Node n_i's time to sleep
T_a	Node n_i's active duration time
T_s	Node n_i's sleep duration time
T_0	The time for transmitting a broadcast message
R	Node's Communication radius
ρ	Network density
N_R	Number of nodes in circle area πR^2, namely $N_R = \pi R^2 \cdot \rho$
λ	Parameter of Poisson distribution for intermittent connectivity
λ_0	$\lambda_0 = N_R / T_a$
α	Parameter for selecting λ that $\lambda = \alpha \cdot \lambda_0$ for intermittent connectivity
P_λ	Broadcast success ratio under parameter λ
t_i^{send}	Node n_i's time to forward broadcast
$T_i^{backoff}$	Node n_i's exponential backoff time
M_i	Node n_i's message buffer
B_i	Node n_i's beacon message
$footer$	Transmitting with DATA to reduce broadcast redundancy
$B_i.Id$	The max Id of the broadcast message received by node n_i
$\Delta T(i)$	The latency of the broadcast message received by node n_i
$\Delta T^k(i)$	Node n_i's latency $\Delta T(i)$ in case k
$t_i^{a,k}$	Node n_i's wake-up time t_i^a in case k

There are two short packets, i.e., *beacon* and *footer*, used in DQSB, as shown in Figure 1. A *beacon* is used by a node to announce its active state when it wakes up. It includes *Id*, *Node_id* and *Wakeup_time*. Field *Id* is used to help its neighbor make forwarding decisions or trigger its neighbors to adjust their sleep scheduling to receive its broadcast message. So, in *beacon*, the value of *Id* will be set to the maximum sequence number of the received broadcast message in its message buffer and it will be updated dynamically. Field *Node_id* and *Wakeup_time* are used to tell neighboring nodes when the node wakes up. The *footer* indicating the transmitting for its neighboring nodes contains fields such as *Forwarder*, *Receivers*, *Message_id* and *End_time*. Suppose that forwarder n_i and n_j will transmit the broadcast message with the same *Message_id* at time t_i^{send} and t_j^{send} respectively ($t_i^{send} < t_j^{send}$). However, if n_j's receivers are contained in n_i's and n_j can hear the *footer* from n_i, then n_j will abort this forwarding task at time t_j^{send} in order to reduce broadcast redundancy. Otherwise, n_j must start its forwarding task with backoff mechanism to avoid a collision when it learns from *End_time* ($End\_time = t_i^{send} + T_0$) in n_i's *footer* that $t_i^{send} < t_j^{send} < End\_time$, indicating n_j's forwarding will take place during n_i's transmitting. Besides, *footer* and *DATA* carried in a packet will be transmitted as a broadcast message in DQSB.

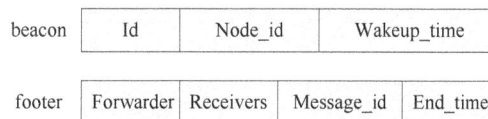

Figure 1. Packet structures of *beacon* and *footer*.

An assumption used in [25] as well is that the time interval between two asynchronous duty-cycled nodes is an exponentially distributed random variable with average T_a/N_R, where T_a is the length of the node's active period after waking up and N_R is the number of nodes in the circle area of communication radius R. A sequence of waking up behavior of nodes is represented by a Poisson process. The probability P_t that more than one node wakes up in a period t is formulated as follows:

$$P_t = 1 - e^{-\frac{N_R}{T_a} \cdot t} - \frac{N_R}{T_a} \cdot t \cdot e^{-\frac{N_R}{T_a} \cdot t} = 1 - (1 + \frac{N_R}{T_a} \cdot t) \cdot e^{-\frac{N_R}{T_a} \cdot t}$$

From the above equation, if the period t is extremely short (approaching zero), P_t approaches zero. Hence, in an asynchronous duty-cycled WSN, the probability that more than one node wakes up simultaneously is almost zero. Consequently, this helps us to ignore the collisions happening among nodes when they wake up and immediately send a short packet. With these assumptions, in asynchronous duty-cycled WSNs, we assume that there is a λ ($\lambda = \alpha \cdot \lambda_0$, $\lambda_0 = N_R/T_a$) such that for all λ' ($\lambda' \geq \lambda$), an evolving graph G(t) is intermittent connectivity during time t, where t $\in$ [0,T] and T is one cycle length that $T = T_a + T_s$. For properly selecting λ to ensure the assumption, we will explore the impact of λ and get the smallest value of α in Section 5.

3.2. Conditions for Successful Broadcast

In an asynchronous duty-cycled WSN, suppose the broadcast time of node n_i is set to t_i^{send} if it is a broadcast forwarder. There is $t_i^{send} = t_i^s - T_0$. n_j receives a message sent by n_i if and only if the duration $T_{i,j}$ for the link $L_{i,j}(t_{i,j}, T_{i,j})$ satisfies the following conditions. As shown in Figure 2, there are two conditions:

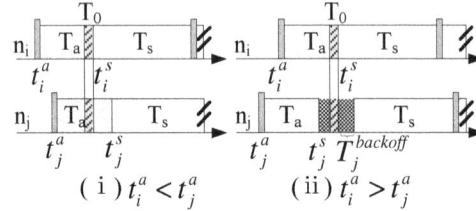

Figure 2. Two conditions of timing relationship between transmitter n_i and its receiver n_j.

(1) If $t_i^a < t_j^a$, link $L_{i,j}$'s duration time $T_{i,j}$ satisfies $T_{i,j} \geq T_0$. In this case, the transmitter n_i wakes up earlier than its receiver n_j. If the link $L_{i,j}(t_{i,j}, T_{i,j})$ between nodes n_i and n_j satisfies the condition, then $t_i^s - t_j^a \geq T_0$, where equation $t_i^s - t_j^a = T_{i,j}$ holds. So, the condition is converted to $T_{i,j} \geq T_0$.

(2) If $t_i^a > t_j^a$, n_j adjusts its sleep schedule by $t_j^s = t_i^s + T_j^{backoff}$ if it cannot receive broadcast messages in an opportunistic way from other nodes. This case indicates that if n_i, serving as the transmitter, wakes up later than the receiver n_j, n_j cannot receive n_i's broadcast message since n_j has already switched to sleep state. Therefore, n_j must prolong its active time and move to sleep state later than the transmitter n_i. It is expressed as $t_j^s = t_j^s + (t_i^a - t_j^a + T_j^{backoff})$, which is further converted to $t_j^s = t_i^s + T_j^{backoff}$ due to $t_j^s - t_j^a = T_a$ and $t_i^a + T_a = t_i^s$. Then, this case is converted to the condition 1.

In summary, all the sleep schedules of nodes need to satisfy the condition 1 in order to let receivers properly receive broadcast messages.

3.2. Quasi-Synchronization Mechanism

A broadcast protocol aims to generate a broadcast tree from the broadcast source node to all the other nodes, and the sleep scheduling relationship between any node and its parent node satisfies the condition that a sender wakes up earlier than its receiver. Quasi-synchronization mechanism proposed in this paper is responsible for this undertaking goal. If a receiver wakes up earlier than its parent node, it is called timing inversion. The mechanism firstly helps nodes determine whether there is a timing inversion in its path of the broadcast tree. If it is, the mechanism requests the related nodes to adjust their sleep schedules. Consequently, the condition remains true among all the nodes in the broadcast tree. All the nodes reach a quasi-synchronized state. Quasi, here, means that all the nodes may not wake up at the same time, but they are able to receive all the broadcast messages sent by the root of the broadcast tree.

The quasi-synchronization mechanism needs to handle 4 cases due to the relationship between a sender and its direct receiver. Suppose that n_i transmits a broadcast message at time t_i^{send}, $t_i^{send} = t_i^s - T_0$ as shown in Figure 3.

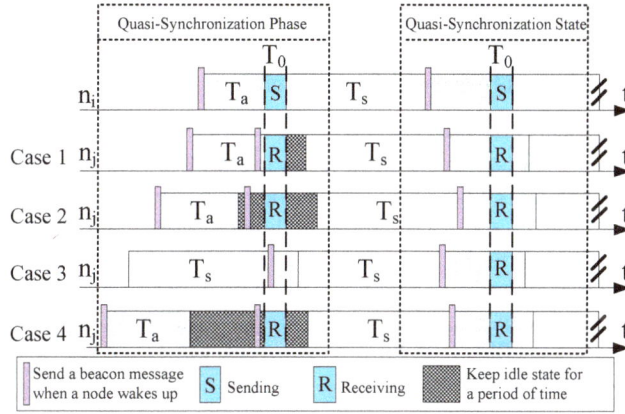

Figure 3. Suppose n_i is a broadcast forwarding node and n_j is its receiver. Different cases are considered in quasi-synchronization mechanism and their adjustment strategies: (Case 1 and Case 2) Early Sleep. (Case 3) Late Wake-up. (Case 4) Isolated Node.

(1) n_j wakes up before n_i. n_j fails to receive n_i's broadcast message. If n_j does not receive the broadcast message from other neighbor nodes during $[t_{n_j,a}, t_{n_j,send} - T_j^{backoff}]$ (case 1) or $[t_{n_j,a}, t_{n_j,s}]$ (case 2), n_j retransmits its *beacon* at time t ($t = t_i^{send} - T_j^{backoff}$) to trigger n_i's forwarding decision. Both case 1 and case 2 are due to receivers' early sleep.
(2) n_j's wake-up happens during n_i's transmitting time. n_j cannot receive n_i's broadcast message correctly. This case is called late wake-up problem and its corresponding solution for n_j is to adjust its next sleep time and sets its $t_{j,s}$ to $t_{j,s} - T_j^{backoff}$.

(3) Case 4 is due to the isolated node problem. When n_j finds that it cannot receive anything else during its active period of $[t_j^a, t_j^s - T_0]$, it prolongs its active time until it receives its neighboring nodes' wake-up message, such as *beacons*. If n_j can learn whether any of their neighbors have the broadcast message it wants from the receiving *beacons*, n_j can decide whether or not to retransmit its *beacon* at time t ($t = t_{n_j,a} + T_j^{backoff}$) to trigger n_i's forwarding and wait for the the upcoming broadcast message. Then, n_j switches to sleep state at $t_{j,s}$ ($t_{j,s} = t_{i,s} + T_j^{backoff}$).

After the adjustments of nodes' sleep schedules in quasi-synchronization in Figure 3, the timing inversion problem is solved.

4. DQSB PROTOCOL DESIGN

4.1. Overview

DQSB aims at giving a solution to multi-hop broadcast and routing in asynchronous duty-cycled WSNs without time synchronization and duty-cycled awareness, and helps nodes to forward broadcast messages and transmit sensed data to a broadcast source node, to reduce broadcast redundancy and keep relative tolerant broadcast latency performance. It is regarded as a joint design approach to achieve reliable broadcast and less latency routing paths for reverse data collection to a broadcast source node. The main idea of the protocol is to let nodes locally develop their own views of the sleep schedule of the whole network. Although in a system viewpoint, the nodes' sleep schedules are not strictly synchronized, the adjustments of nodes' sleep schedules are good enough to guarantee that all the nodes receive broadcast messages. Under DQSB, a data collection node, such as sink, periodically broadcasts messages to inform other nodes to transmit their sensing data to it. The other nodes firstly try to receive broadcast messages in an opportunistic way. If this fails, they will trigger one of their neighbor nodes' forwarding decisions and adjust their duty-cycle in a local and distributed manner for receiving. As a result, nodes' sleep schedules are coordinated. Every node learns its next hop to the sink by a minimal latency path due to no or less waiting time and no synchronization delay.

4.2. Distributed Quasi-Synchronized Broadcast

DQSB is composed of two basic components: (1) Forwarding decision, which helps nodes to know whether or not to forward received broadcast messages; and (2) Adaptive sleep scheduling adjustment, which is triggered when a node is aware of the upcoming transmissions which it misses. The node will adjust its sleep schedule in order to receive necessary messages. DQSB has five tasks to complete: forwarding decision, sending task abort, managing early sleep, tackling late wake-up and dealing with isolated nodes. The protocol is presented in Algorithm 1, where variables at each node have been described in Table 1. For consistency with Figure 2 and Figure 3, let's illustrate how n_j finishes the following tasks with one of its neighboring nodes, such as n_i.

When a node n_j wakes up, at first it drops expired messages and gets the newly received broadcast messages in its message buffer (line 1-2). Then, it transmits its *beacon* as a one-hop broadcast immediately, indicating the maximum identification of received broadcast messages. If the message buffer is empty, the value of *Id* in the *beacon* is set to -1 (line 3-5). After transmitting its *beacon* B_j, it will receive *beacons* from its neighboring nodes during the interval of $[t_j^a, t_j^s - T_0]$, such as B_i from neighboring n_i.(line 6-30).

Task 1: *Forwarding Decision (line 7-10).* Forwarding decision is made by n_j according to the received *beacons* from neighboring nodes during n_j's active interval of $[t_j^a, t_j^s - T_0]$. n_j's sending is triggered if and only if the condition $(M_j != null) \&\&(\exists m \in M_j, m.Id > B_i.Id, B_i \in beacons)$ is satisfied. If n_j really forwards a message, its sending time t_j^{send} is set to $t_j^s - T_0$.

Task 2: *Sending Task Abort (line 22-24).* Sending task abort happens if and only if node n_j listens to the channel and receives the *footer* before t_j^{send}, and the *footer* indicating the transmitting for its neighbor nodes at time t_j^{send} has been done.

Task 3: *Managing Early Sleep (line 11-19).* Early sleep occurs if and only if n_j receives more than one *beacon* during its active period of $[(t_j^a, t_j^s - T_0)]$ and meets one of the following conditions: (1) $\forall m \in M_j, m.Id < B_i.Id, B_i \in beacons$ or (2) $M_j == null \&\& B_i.Id \neq -1, B_i \in beacons$. According to the analysis of Figure 2, the relationship between transmitter n_i and its receiver n_j satisfies the condition $t_i^a > t_j^a$. Case 1 and 2 are recognized as the early sleep problems discussed in Figure 3. Their solutions to trigger neighbors' transmitting are given, respectively: (1) If $(T_a - T_0 < T_{i,j} < T_a)$, n_j updates its $B_j.Id$ when receiving broadcast messages during $[t_j^a, t_i^{send} - T_j^{backoff}]$ (Case 1); (2) If $(0 < T_{i,j} < T_a - T_0)$, n_j updates its $B_j.Id$ when receiving broadcast messages during $[t_j^a, t_j^s]$ (Case 2). After updating its $B_j.Id$ during these times, if the condition $B_j.Id < B_i.Id$ is still satisfied, n_j retransmits its beacon B_j at time $t = t_i^{send} - T_j^{backoff}$ to trigger n_i's forwarding. Then, n_j only waits for the upcoming broadcast message. After receiving, n_j adjusts its t_j^s ($t_j^s = t_i^s + T_j^{backoff}$), namely condition $t_i^a < t_j^a$ holds in the following duty cycle.

Algorithm 1: Distributed quasi-synchronized broadcast

begin
  if(node n_j's wakeup time t_j^a arrive) then
    Move to idle state;
    $M_j \leftarrow getMessageBuffer()$;
    if($M_j != null$) then $B_j.Id \leftarrow M_j.getMaxId()$;
    else $B_j.Id \leftarrow -1$;
    Transmit its beacon B_j;
    While ($t_j^a \leq t \leq t_j^s - T_0$) do
      if(receive beacon B_i from n_i) then
        if($(M_j != null)$ & &($\exists m \in M_j.m.Id > B_i.Id$)) then
          $t_j^{send} \leftarrow t_i^s - T_0$;
          Wait t_j^{send} arrive and broadcast m;
        else if($\forall m \in M_j.m.Id < B_i.Id$) || ($M_j == null$ & & $B_i.Id \neq -1$)) then
          if($T_a - T_0 < T_{i,i} < T_a$) then
            Update $B_j.Id$ when receives broadcast message during $t_j^a < t < t_i^{send} - T_j^{backoff}$;
          else if($0 < T_{i,j} < T_a - T_0$) then
            Update $B_j.Id$ when receives broadcast message during $t_j^a < t < t_j^s$;
          if($B_j.Id < B_i.Id$) then
            $t_j^s \leftarrow t_i^s + T_j^{backoff}$;
            Retransmit its beacon B_j at time t ($t = t_i^{send} - T_j^{backoff}$) to trigger n_i's forwarding;
            Wait and receive broadcast message;
      else if(neighbor n_i is transmitting && $t_i^{send} < t_j^a < t_i^s$) then
        $t_j^s \leftarrow t_j^s - T_j^{backoff}$;
      else if(n_j receives *footer* before t_j^{send}) then
        if(The footer indicates the transmitting for its neighbor nodes has been done) then
          Abort its sending at t_j^{send};
      else if(n_j can't receive any message) then
        Prolong its active time until it receives neighbor nodes' *beacons*;
        if($B_i.Id > B_j.Id, B_i.Id \in beacons$) then
          Retransmit its beacon B_j at time t ($t = t_i^a + T_j^{backoff}$) to trigger n_i's forwarding;
          Wait and receive broadcast message;
          $t_j^s \leftarrow t_i^s + T_j^{backoff}$
  if(node j's sleep time t_j^s arrive) then
    Move to sleep state
end

Task 4: *Tackling Late Wake-up (line 20-21).* As case 3 shows in Figure 3, a late wake-up occurs if and only if n_j listens to the channel and overhears neighboring node n_i's transmitting during $[(t_j^a, t_j^s - T_0)]$ and n_j's wake-up time t_j^a satisfies condition $t_i^a < t_i^{send} < t_j^a < t_i^s$. However, here, n_i is the transmitter, and the condition $T_{i,j} < T_0$ cannot satisfy the condition $T_{i,j} \geq T_0$ required by Condition (1) in Figure 2. The corresponding solution for n_j is to set its $t_{j,s}$ to $t_{j,s} = t_{j,s} - T_j^{backoff}$.

Task 5: *Dealing With Isolated Nodes (line 25-30).* Isolated nodes occur if and only if n_j receives no *beacon* and listens nothing else during its active period of $[t_j^a, t_j^s - T_0)]$. Although this problem may not happen at the initial time of a WSN due to our assumption, it may really occur if nodes recover from failure or join in the network afterwards. As case 4 shows in Figure 3, when n_j finds that it cannot receive any *beacon* message and anything else during its active period of $[t_j^a, t_j^s - T_0]$, it prolongs active time until it receives neighboring nodes' *beacons* e.g. B_i. If the condition that $B_i.Id > B_j.Id, B_i.Id \in beacons$ holds, n_j retransmits its beacon B_j at time t ($t = t_{i,a} + T_i^{backoff}$) to trigger n_i's forwarding and waits for the upcoming broadcast message. n_j switches to sleep state at $t_{j,s}$ ($t_{j,s} = t_{i,s} + T_j^{backoff}$).

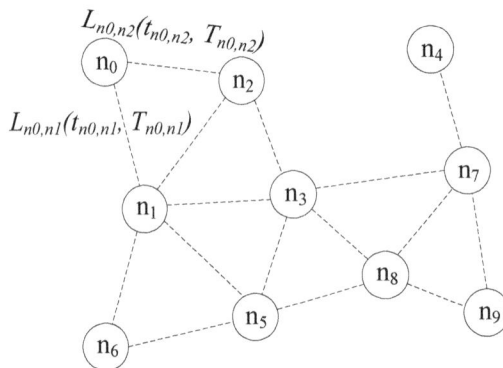

Figure 4. An evolving graph *G(t)* of ten nodes in an asynchronous duty-cycled WSN. *G(t)* = (*N, E(t))*, where $N = \{ n_0, n_1, \cdots, n_9 \}$ and the dotted lines denote intermittent connective links $L_{i,j}(t_{i,j}, T_{i,j})$ between n_i and n_j with $L_{i,j}(t_{i,j}, T_{i,j}) \in E(t)$.

We further discuss DQSB in detail with an example. As shown in Figure 4, suppose there are ten nodes in an asynchronous duty-cycled WSN. n_0 is in charge of the data collection and periodically broadcasts some DATA packets to the other nodes. Figure 5 gives an overview of the operation sequences of DQSB regarding the scenario in Figure 4. In the forwarding decision phase, because n_0 is the broadcast source at first, $B_0.Id$ is set to 0 for the first broadcast message and is transmitted when it wakes up (Sequence 1). As n_0's neighbors, both of n_1 and n_2 wake up before n_0, they can receive n_0's beacon message B_0 and fail to receive n_i's broadcast message because of early sleep problem. But when they can receive the broadcast message from other neighboring nodes, they will update their *Id* values of the beacons during $[t_1^a, t_1^s]$ and $[t_2^a, t_0^{send} - T_2^{backoff}]$ respectively. Otherwise, $M_1 == null$, $M_2 == null$ and $B_0.Id \neq -1, B_0 \in beacons$ satisfy the condition (2) in Task3. Here, both $B_1.Id$ and $B_2.Id$ will be set to -1. Then they retransmit their *beacons* before t_0^{send} (Sequence 2 and 3). When n_0 receives B_1 and B_2 from n_1 and n_2 respectively, any of them will trigger n_0's sending at t_0^{send} due to $B_0.Id > B_1.Id$ or $B_0.Id > B_2.Id$ (Sequence 4). When n_1 and n_2 receive the broadcast messages, they move to sleep state at time t_1^s ($t_1^s = t_0^s + T_1^{backoff}$) and t_2^s ($t_2^s = t_0^s + T_2^{backoff}$) respectively (Sequence 5 and 6). Suppose n_3 finds it cannot receive any *beacon* from neighboring nodes during its active period of $[t_3^a, t_3^s - T_0]$, it prolongs its active time till receiving neighboring nodes' *beacons*, e.g., n_1 and n_2. Then n_3 retransmits B_3 because of $B_1.Id > B_3.Id$ or $B_2.Id > B_3.Id$ (Sequence 7). After receiving n_3's beacon B_3, both n_1 and n_2 will make a forwarding decision (Sequence 8 and 9). But in asynchronous duty-cycled WSNs, n_1 and n_2 do not transmit simultaneously according to their expected sending time. Since n_2 listens n_1's transmitting and learns from the *footer* indicating the broadcast message n_3 wants has been transmitted by n_1, n_2 will abort the sending task at time t_2^{send} to avoid collision and reduce broadcast redundancy (Sequence 10). Unfortunately n_6's wake-up happens during n_1's transmitting time, n_6 cannot receive the broadcast message correctly. Then n_6 must set its sleep time $t_{6,s}$ to $t_{6,s} - T_6^{backoff}$ (Sequence 11). Upon this adjustment, n_6 will wake up at time t_6^a ($t_1^a < t_6^a < t_1^{send}$) in next duty cycle. The worse case for n_6 is that it cannot receive anything else during its active period from its neighbors such as n_1 and n_5. It will prolong its active time till receiving neighboring nodes' *beacons*, i.e. B_1 from n_1. If $B_1.Id > B_6.Id$ holds, n_6 will retransmit its beacon B_6 at time t ($t = t_{n_1,a} + T_6^{backoff}$) to trigger n_1's forwarding. n_6 switch to sleep state at $t_{6,s}$ ($t_{6,s} = t_{1,s} + T_6^{backoff}$).

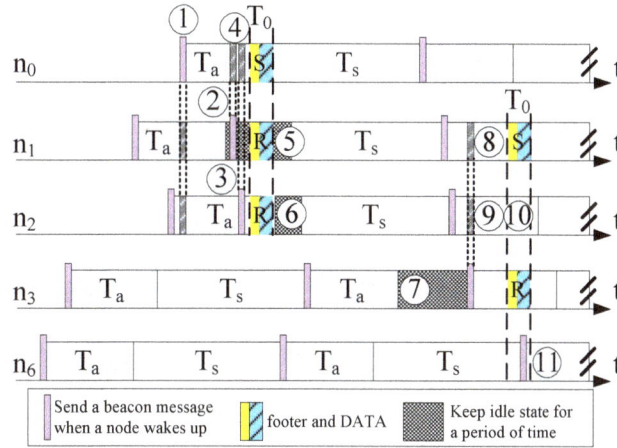

Figure 5. Overview of distributed quasi-synchronized broadcast protocol of DQSB regarding the scenario in Figure 4. Every node transmits a *beacon* message when it wakes up at time t_i^a, indicating the maximum broadcast message ID it recently receives. Broadcast source node, such as n_0, makes broadcast forwarding decision according to the received *beacons*. If so, transmit at $t_i^{send} = t_i^s - T_0$, where t_i^s is node n_i's time to switch to sleep state. Other nodes with received *beacons* will adjust their sleep scheduling and trigger their neighbor nodes to forward based on our quasi-synchronized mechanism in Figure 3, i.e. the adjustment of n_1 and n_2 for n_0's broadcast due to early sleep problem.

4.3. State Diagram Description

DQSB protocol has 7 states. They are idle, sleep, forward-decision, receiving, routing, transmitting and forward-unicast. Let's revisit the example shown in Figure 4. Figure 6 illustrates the state transition diagram for DQSB's running triggered by different conditions given in Table 2.

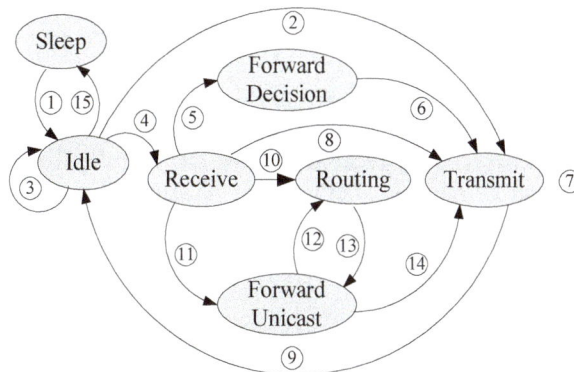

Figure 6. The state transition diagram for DQSB.

Table 2. Conditions for DQSB's state transition diagram.

Id	Transition Condition
1	node n_i's wakeup time t_i^a arrive;
2	node n_i transmit its beacon B_i at time t_i^a;
3	node n_i's active time t, $t_i^a \le t \le t_i^s - T_0$;
4	node n_i receives *beacon, broadcast* or *unicast* message;
5	node n_i receives *beacon*, such as B_j from node n_j;
6	node n_i satisfies $(M_i \mathrel{!=} null) \,\&\&(\exists m \in M_i, m.Id > B_j.Id)$;
7	node n_i's forwarding broadcast time $t_i^{send} = t_i^s - T_0$;
8	node n_i abort the broadcasting task at t_i^{send} by received *footer*;
9	node n_i finishes transmitting task and move to idle state;
10	node n_i learns the next hop from received *broadcast* message;
11	node n_i receives *unicast* message or *sensed data* from itself;
12	node n_i looks up the route table for forwarding unicast message;
13	node n_i gets the next hop for *unicast* message from route table;
14	node n_i transmits the *unicast* message when the channel is idle;
15	node n_i's sleep time t_i^s arrive;

1. Suppose n_0, as a sink, broadcasts and triggers the other nodes to transmit their sensed data to it. When *Condition 1* is established for n_0, n_0 switches to *idle* state and transmits a beacon B_0 with $B_0.Id = 0$ (*Condition 2*). Under *Condition 3*, n_0 is waiting for *beacons* from its neighbors to make forwarding decision. But n_1 and n_2 wake up before n_0, they can receive n_0's beacon message B_0 and fail to receive n_i's broadcast message because of early sleep problem. Both $B_1.Id$ and $B_2.Id$ are equal to -1, $B_1.Id < B_0.Id$ and $B_2.Id < B_0.Id$. They will retransmit their *beacons* before t_0^{send} which will be received by n_0 (*Condition 4*).

2. With the received beacons B_1 and B_2, n_0 makes a forwarding decision (*Condition 5*). Either the value of $B_1.Id$ or $B_2.Id$ holds by *Condition 6*. Then, n_0 sets forwarding broadcast task at time t_0^{send} and waits for transmitting (*Condition 7*). After its sending, n_1 and n_2 receive the broadcast message (*Condition 4*), and $B_1.Id$ and $B_2.Id$ are set to 0 when they wake up next time to transmit their *beacons*. If it is the sleep time of $t_{0,s}$ for n_0, n_0 turns into *sleep* state (*Condition 15*).

3. For n_1 and n_2, they learn the next hop to n_0 from received broadcast message and switch to *routing* state (*Condition 10*). In the next duty cycle, when their sensed data are delivered by the upper layer (*Condition 11*), they wake up and transmit their beacons according the routing table immediately if the channel is idle (*Condition 12, 13 and 14*). If there is no message to forward, then switch to *idle* state (*Condition 9*).

4. When n_3 and n_6 wake up, they perform in the same way as n_1 and n_2 in step (1). Here, it is important to notice both n_1 and n_2 receive beacons B_3 and B_6 from n_3 and n_6. Due to the case that $B_3.Id = -1$ and $B_6.Id = -1$, both n_1 and n_2 launch forwarding task at time t_1^{send} and t_2^{send}, respectively. Because t_1^{send} is earlier than t_2^{send} according to the forwarding mechanism, it is the right time for *footer* to let n_2 abort its forwarding task (*Condition 8*) and further reduce broadcast times for energy efficiency.

5. After receiving the broadcast message, n_3 and n_6 execute what are described in step (3) to transmit their sensed data to n_0.

4.4. Further Discussion

Property 1 (Validity and Reliability) *With our assumption of intermittent connectivity , for any node n_j, there is at least one node of its neighborhoods that have received the broadcast message n_j wants. Then n_j finally receives the broadcast message from one of its neighbor n_i within $\Delta T(j)$,*

$$\Delta T(j) \le \Delta T(i) + t_i^s - t_j^a + T, where\ T = T_a + T_s.$$

Proof: This property is proved based on the cases shown in Figure 3. Our assumption of intermittent connectivity implies case 4 that node receives no beacon and listens nothing else will never exist. But we introduce this case to help nodes recover from failure or join in the network, which can improve the robustness and adaptively for our protocol. We will prove this in Property 2. So, except case 4, we will deduce the latency ΔT_1, ΔT_2 and ΔT_3 for case 1, case 2 and case 3 in Figure 3, respectively. According to the assumption, we can infer a significant result that every new broadcast message can arrive at one of n_j's neighbor nodes such as node n_i, then this can make sure n_j's successful receiving for the broadcast message from n_i, even if n_j cannot receive the broadcast messages during time t, $t_{j,a} < t < t_{i,send} - T_j^{backoff}$. Because $B_j.Id$ still satisfies inequality $B_j.Id < B_i.Id$ which will trigger n_i's forwarding decision. For case 1, the maximum latency of node n_j's receiving for the broadcast message equals to $\Delta T^1(j)$, $\Delta T^1(j) = \Delta T(i) + t_i^s - t_j^{a,1}$. Similar to case 1, the maximum latency of case 2 satisfies the equation that $\Delta T^2(j) = \Delta T(i) + t_i^s - t_j^{a,2}$, and $\Delta T^1(j) < \Delta T^2(j)$ because of $t_j^{a,1} > t_j^{a,2}$. Case 3 is different from case 1 and case 2 for its late wake-up feature mentioned in Task 4. Obviously, this case will lead to longer latency than that can be obtained by the expression that $\Delta T^3(j) = \Delta T(i) + t_i^s - t_j^{a,3} + T$. Since $\Delta T^3(j) > \Delta T^2(j) > \Delta T^1(j)$, the maximum latency of node n_j's receiving the broadcast message $\Delta T(j)$ satisfies the inequation that $\Delta T(j) \le \Delta T(i) + t_i^s - t_j^a + T$, then Property 1 holds.

Property 2 (Robustness and Adaptively) *When n_j recovers from failure or newly joins in a network, there is at least one node of its neighborhoods that have received the broadcast message n_j wants. Then n_j finally receives the broadcast message from one of its neighbor n_i within $\Delta T(j)$,*

$$\Delta T(j) \le \Delta T(i) + T,\ where\ T = T_a + T_s.$$

Proof: To make DQSB be tolerant with node's failure in realistic environment and be adaptive to new node's joining, we consider the case 4 shown in Figure 3 that n_j receives no *beacon* and nothing else during its active period of $[(t_j^a, t_j^s - T_0)]$. Let $\Delta T^4(j)$ be receiving latency of the broadcast message in case 4. Then, $\Delta T^4(j) = \Delta T(i) + t_i^a - t_j^a + T_a$. In the most extreme, the maximum latency occurs when node n_j's wakeup time t_j^a happen in node n_i's sleep time t_i^s, and node n_i's next wakeup time t_i^a satisfies the condition that $t_i^a - t_j^a = T_s$. This can make the equation of $\Delta T^4(j)$ convert to $\Delta T^4(j) = \Delta T(i) + T_s + T_a$, then Property 2 has been proved upon the condition $T = T_s + T_a$.

With Property 1 and Property 2, the lower bound of latency for any node is given.

4.4. Less Latency Routing

We observe that a routing protocol to a sink node is related to a broadcast protocol. If a broadcast protocol constructs a good bottom-up tree path to all the other nodes in a WSN, the tree path in the reverse direction, then is a satisfying road map for all the nodes transmitting their data to the sink node. Without time synchronization in asynchronous duty-cycled WSNs, unicast protocols, such as B-MAC, WiseMAC and X-MAC, have more or less waiting latency because of the unawareness of sleep schedules of neighboring nodes. DQSB's quasi-synchronization helps the nodes to properly adjust others' wake-up schedules. For instance, n_6 wakes up earlier than n_1, and n_1 is earlier than n_0 in Figure 5. Although they do not wake up at the same time, if n_6 and n_1 adjust their sleep schedules according to quasi-synchronization mechanism, the broadcast messages sent by n_0 are received by all its neighbor nodes in one cycle, such as n_1 and n_2. The result of their sleep schedule relationship comes out that n_0 wakes up earlier than n_1, and n_1 is earlier than n_6. So when n_6 and n_1 wake up respectively, there is no extra waiting time for them to transmit their sensed data to the sink node n_0. Therefore, the less latency route for each sensed data from source to data collection node is constructed, as shown in Figure 7. The broadcast forwarding for each broadcast message only needs four nodes, i.e., n_0, n_1, n_3 and n_7. This is of great significance for energy efficient in terms of transmitting times, rather than every node's forwarding. In addition, the less latency routes are learned during broadcasting, such as $n_6 \rightarrow n_1 \rightarrow n_0$, $n_5 \rightarrow n_3 \rightarrow n_1 \rightarrow n_0$, $n_9 \rightarrow n_7 \rightarrow n_3 \rightarrow n_1 \rightarrow n_0$ and so on.

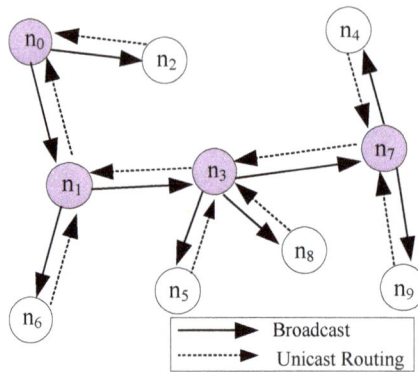

Figure 7.   Every node's less latency routing path to the broadcast source node with the distributed quasi-synchronized broadcast in our DQSB protocol for the simple network scenario shown in Figure 4.

6. SIMULATION AND ANALYSIS

6.1. Simulation Setting

The ONE simulator is an open source tool for Delay-tolerant Networking (DTN), specifically designed for evaluating routing and application protocols in intermittent connective networks, such as asynchronous duty-cycled WSNs. We develop our simulator based on the ONE simulator [11] to evaluate our DQSB protocol. To satisfy the evaluation requirements, we develop extensive simulator functions based on the ONE simulator as shown in Figure 8. *Broadcast Message Event Generator* is used to generate broadcast messages in the given interval; *Random Sleep Scheduling Generator* lets randomly deployed nodes work in asynchronous duty-cycled WSNs; *Reliable Broadcast* provides distributed quasi-synchronized broadcast; and *DQSB* is applied to data collection based on *Reliable Broadcast*. We also implement Hybrid-cast and OppFlooding protocols in order to compare with our DQSB protocol.

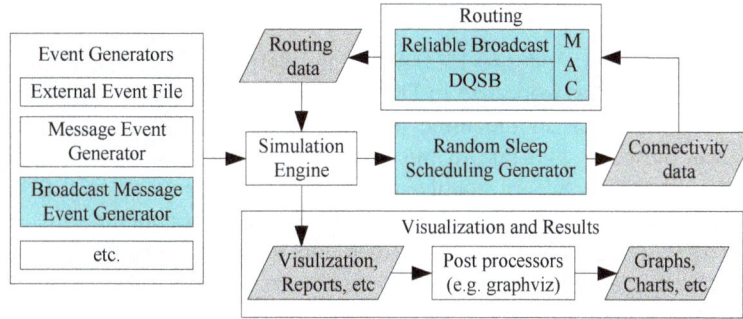

Figure 8. An Extended Simulator based on The ONE Simulator

6.2. Regarding Duty Cycles

We evaluate the performance in asynchronous duty-cycled WSNs with various duty cycles. In this simulation, wireless loss rate is set to 0.1, wireless communication range to 15m and transmitting speed to 250kbps. The size of a broadcast message is fixed as 512 bytes and its transmitting time of T_0 is 50ms. We randomly generate 10 topologies with 200 nodes, and run on each topology for 10 times.

Figure 9 illustrates the performance of forwarding times and broadcast latency, respectively. From Figure 9(a), we notice that DQSB outperforms Hybrid-cast and OppFlooding. This is because node's forwarding is triggered by its receivers in DQSB. The nodes that cannot receive the beacon will adjust their sleep schedules in order to receive broadcast messages, which is different from the other two protocols. Regarding broadcast latency in Figure 9(b), DQSB behaves particularly because its latency does not decrease in spite of the increasing of duty cycle. This contributes to DQSB's mechanism that each node in one cycle either receives a broadcast message or forwards the message received in the last cycle. Broadcast latency is related to duty cycle and forwarding times. The relationship between them is that one-hop latency follows the increasing of duty cycle in that a node launches forwarding at $t_i^{send} = t_i^s - T_0$. Generally speaking, the more nodes receive a broadcast message during one forwarding, the fewer forwarding times is. Consequently, broadcast latency in DQSB is a tradeoff between duty cycle and forwarding times.

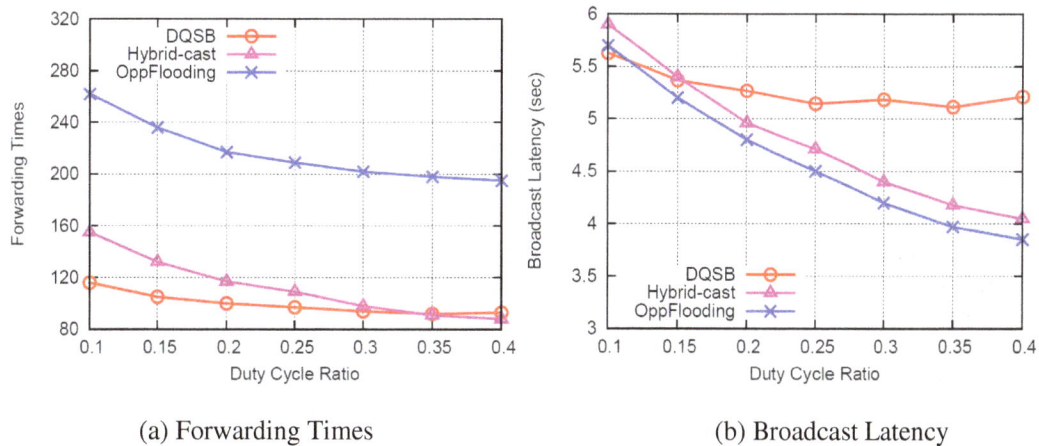

(a) Forwarding Times

(b) Broadcast Latency

Figure 9. Forwarding Times and Broadcast Latency with various duty cycles.

6.3. Regarding Network Size

Network size varies from 200 nodes to 1200 nodes and duty cycle is set to 0.2. The experiment aims to show the impact of network size in DQSB. As shown in Figure 10(a), as network size goes up, forwarding times of all the three protocols exhibit an increasing trend. However, DQSB outperforms the other two due to the same reason given before. As shown in Figure 10(b), compared with other two protocols, the broadcast latency of DQSB keeps relative tolerant and stable as the increasing in network size. So, we can conclude that DQSB can be recognized as a tradeoff between broadcast times and broadcast latency. One the one hand, node's forwarding is triggered by its receivers which helps to reduce forwarding times. On the other hand, only one of the nodes which receive beacons from neighbors forwards a broadcast message (it is an early sleep node), and the late wake-up nodes are exempt from forwarding these messages. Furthermore, quasi-synchronization greatly helps to let more nodes receive broadcast messages at each duty cycle and hence reduce forwarding times. Consequently, when network size expands, the number of nodes which can receive broadcast messages also goes up. The necessary forwarding times remains stable.

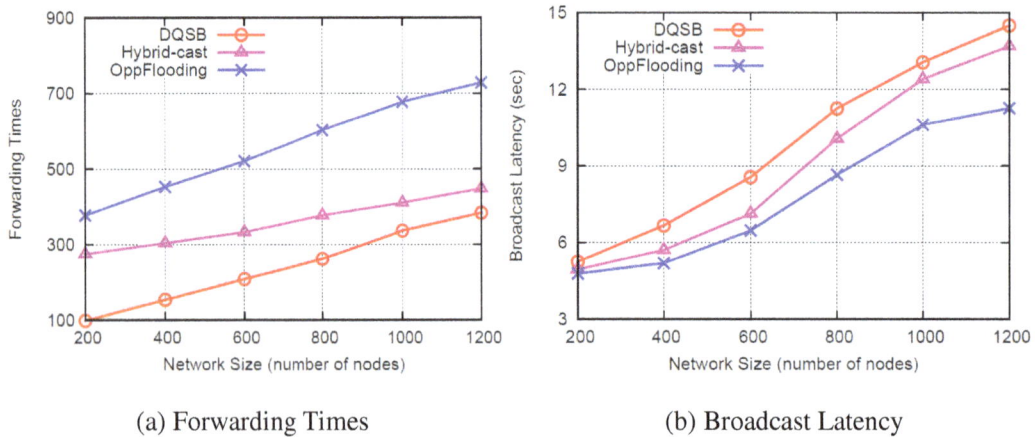

(a) Forwarding Times   (b) Broadcast Latency

Figure 10. Forwarding Times and Broadcast Latency with various network sizes.

6.3. Regarding Reliability with Unreliable Links

In Hybrid-cast and OppFlooding protocols, unreliable links which result in packet loss is not clearly discussed. In our DQSB protocol, *beacon* packets are sent immediately as nodes wake up. These packets are dual-folded. (1) They trigger broadcast forwarding. When receiving beacons, nodes are able to decide whether or not to forward the received broadcast messages; (2) They facilitate DQSB tolerate unreliable links. As we known, wireless links are not always reliable for many reasons in realistic environment, especially in wireless sensor networks. Consider a scenario that some nodes applying DQSB do not receive the broadcast message when a triggered node forwards a broadcast message. This does not matter because n_i's failure lets itself keep the value of $B_i.Id$ in its beacon B_i and it triggers its neighbors's forwarding. Thus, confronting an unreliable link, n_i is able to receive a specific message if one of its neighbors receives it. The performance of reliability under unreliable links is shown in Figure 11. Regarding forwarding times, DQSB performs better than the other two protocols under the environment of unreliable links with wireless link loss rate equals to 0.1 shown in Figure 9. Even with loss rate of 0.3, broadcast latency of DQSB is still acceptable with the value of about 6 seconds.

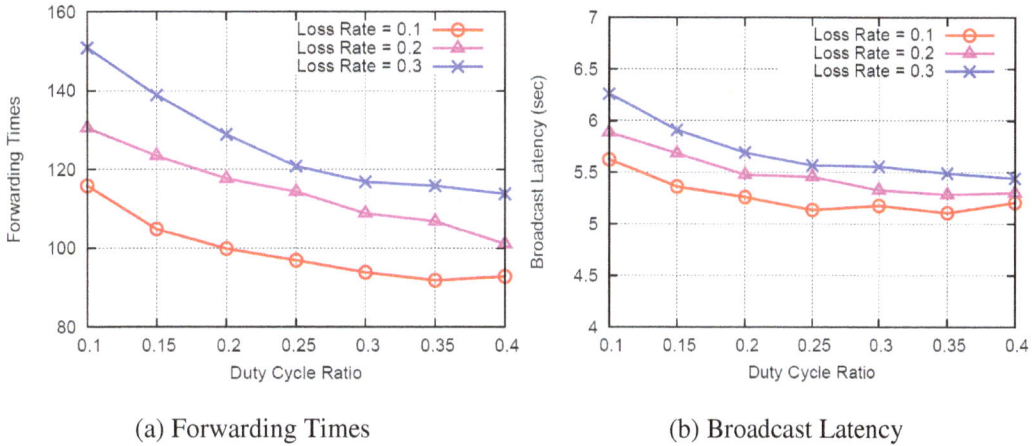

(a) Forwarding Times   (b) Broadcast Latency

Figure 11. Forwarding Times and Broadcast Latency with various link loss rates.

6.5. Regarding Less Latency Routing

Different from the existing multi-hop broadcast protocols in asynchronous duty-cycled WSNs, DQSB is a joint design for reliable broadcast and less latency routing paths for reverse data collection to broadcast source node, such as a sink. In this simulation, a sink node informs other nodes to send their sensed data for data collection, and sensed data are transmitted along the less latency routing paths learned by quasi-synchronized mechanism in DQSB protocol. Packet size is set to 256 bytes for sensed data and packet generating interval is [25, 35] seconds. A sink which broadcasts a message helps other nodes learn their paths from themselves to the sink when they receive the broadcast message. Each node in the network is able to complete this task because of DQSB's reliability for broadcast explained before. Figure 12 shows that DQSB behaves better in latency than LPL (Low Power Listening) which is simple and asynchronous, and adopts long preamble to make the receiver keep awake for a period of time to receive the data. So, the latency is due to the waiting time for both sender and receiver. DQSB solves this problem depending on reliable broadcast and its quasi-synchronized sleep scheduling. Figure 12 shows the less average latency for each hop in DQSB. It also illustrates that average latency for each hop in DQSB is not influenced by duty cycle due to no or less waiting time introduced.

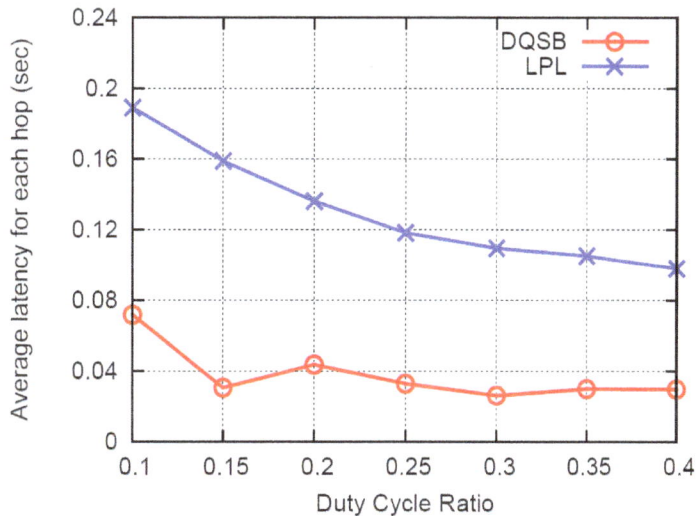

Figure 12. Average latency for each hop.

6.6. Selecting λ for Intermittent Connectivity

Parameter λ_0 indicates the number of nodes that are waken at t. According to Poisson distribution, it is proportional to N_R and inversely proportional to T_a. For any node n_i in a network, if network density increases, n_i's number of neighborhood N_R increases. Thus, n_i's probability to reach other nodes via its neighbor nodes may increase. Moreover, if T_a increases, n_i's probability to be in connection with its neighborhood increases because of the increment of node's duty cycle. But the number of neighbor nodes decreases if λ_0 declines.

We believe that if duty-cycle scheme for nodes is introduced to a connected network, the network exhibits its intermittent connection feature. We should choose the proper value of α, and let $\lambda = \alpha \cdot \lambda_0$ in order to guarantee DQSB's feasibility, i.e., there is at least one path for any node in the network to other nodes via some intermediate nodes within a period (e.g., one cycle length T).

Table 3. Least value of α for $P_\lambda = 1.0$.

Network Size	T_a	λ ($P_\lambda = 1.0$)	Least value of α
200 nodes	0.1	50	0.3808
200 nodes	0.2	40	0.6093
400 nodes	0.1	40	0.1467
400 nodes	0.2	30	0.2201
600 nodes	0.1	30	0.0725
600 nodes	0.2	20	0.0966

In this set of simulations, we investigate the impact of the parameter λ and select a reasonable α to satisfy the assumption. This implies that if this assumption holds, every node in the network receives broadcast messages due to Property 1 and Property 2 of DQSB. So, here we use broadcast success ratio P_λ instead of network's intermittent connectivity to represent whether a network is connected. Figure 13 shows the change of P_λ with λ which varies from 0 to 80 in different network sizes when $T_a = 0.1$ and $T_a = 0.2$, respectively. For a given λ, P_λ increases with network size. When $\lambda = 20$, if network size changes from 200 nodes to 600 nodes, P_λ goes up from 0.5075 to 0.9215 in Figure 13(a) and from 0.8861 to 1.0 in Figure 13(b), respectively. Comparing Figure 13(a) with Figure 13(b), we observe that under the same λ and network size, if T_a moves larger, P_λ also increases. When $\lambda = 20$ and network size is 200 nodes, $P_\lambda = 0.5075$ with $T_a = 0.1$ while $P_\lambda = 0.8861$ with $T_a = 0.2$. Therefore, we set λ to $P_\lambda = 1.0$, and compute the least value of α in terms of the equation $\lambda = \alpha \cdot \lambda_0$, where λ_0 is available by N_R and T_a. Least values of α ave given in Table 3.

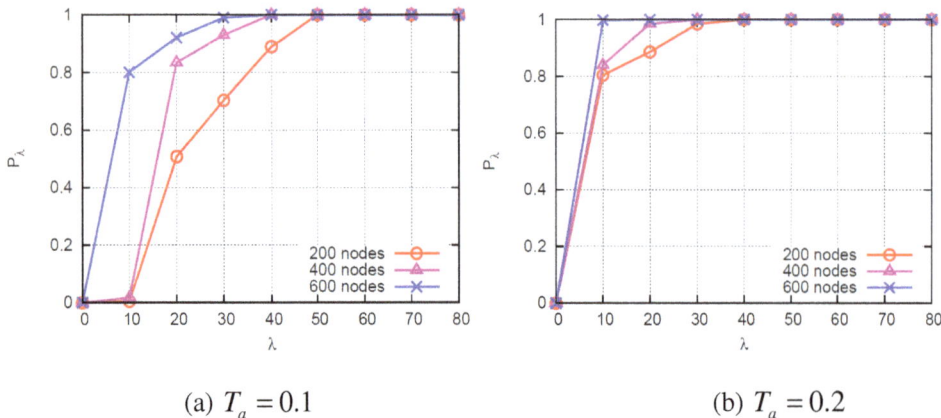

(a) $T_a = 0.1$ (b) $T_a = 0.2$

Figure 13. The impact of the parameter λ under $T_a = 0.1$ and $T_a = 0.2$.

7. CONCLUSIONS

In this paper, we neither assume time synchronization which requires all neighboring nodes wake up at the same time, nor assume duty-cycled awareness which makes it difficult to use in asynchronous WSNs. A reliable broadcast protocol called DQSB is proposed by a distributed and quasi-synchronized manner for duty-cycled WSNs. Quasi-synchronization is reached after nodes execute DQSB in a local and distributed way. Under DQSB, a sink periodically broadcast. After receiving the broadcast messages, other nodes can build their paths to the sink for transmitting their sensed data. Moreover, these paths exhibit less latency because of no or very little waiting time. Simulation results show that DQSB performs well in broadcast times and keep relative tolerant broadcast latency performance. DQSB can be recognized as a tradeoff between broadcast times and broadcast latency. Further, it is still feasible under unreliable links. Our future work is to focus on applying DQSB to real WSN platforms, e.g., micaz and telosb, and investigate its performance.

ACKNOWLEDGEMENTS

This research work is partially supported by the Natural Science Foundation of China under grant No. 60973122, the 973 Program in China under grant No. 2009CB320705, and 863 Hi-Tech Program in China under grant No. 2011AA040502. We thank all of the anonymous reviewers of this work for their valuable comments.

REFERENCES

[1]   J. Yick, B. Mukherjee, and D. Ghosal, "Wireless sensor network survey," *Computer Networks*, vol. 52, no. 12, pp. 2292–2330, August 2008.

[2]   Y. Wang, K. Li, and J. Wu, "Distance Estimation by Constructing The Virtual Ruler in Anisotropic Sensor Networks," *in Proc. of IEEE INFOCOM*, 2010.

[3]   Dr.B.Vinayaga Sundaram, G. Rajesh, A. Khaja Muhaiyadeen, and et al, "A Simple Priority-Based Scheme for Delay-Sensitive Data Transmission Over Wireless Sensor Networks," *International Journal of Wireless & Mobile Networks* (IJWMN), vol. 4, no. 1, February 2012.

[4]   F. Safaei, H. Mahzoon, and M.S. Talebi, "An Efficient Approach for Data Aggregation Routing using Survival Analysis in Wireless Sensor Network," *International Journal of Wireless & Mobile Networks* (IJWMN) vol. 3, no. 2, April 2011.

[5]   J. Li and G. AlRegib, "Network Lifetime Maximizationfor Estimation in Multihop Wireless Sensor Networks," *IEEE Transactions on Signal Processing*, vol. 57, no. 7,pp. 2456–2466, June 2009.

[6]   F. Liu, C.-Y. Tsui, and Y. J. Zhang, "Joint Routing and Sleep Scheduling for Lifetime Maximization of Wireless Sensor Networks," *IEEE Transactions on Wireless Communications*, vol. 9, no. 7, pp. 2258–2267, 2010.

[7]   A. R. Swain, R. C. Hansdah, and V. K. Chouhan, "An Energy Aware Routing Protocol with Sleep Scheduling for Wireless Sensor Networks," *in Proc. of the 24th IEEE International Conference on Advanced Information Networking and Applications*, pp. 933–940, 2010.

[8]   J. Polastre, J. Hill, and D. Culler, "Versatile Low Power Media Access for Wireless Sensor Networks," *ACM SenSys*, pp. 95–107, 2004.

[9]   A. El-Hoiydi and J.-D. Decotignie, "WiseMAC: An Ultra Low Power MAC Protocol for the Downlink of Infrastructure Wireless Sensor Networks," *in Proc. Of ISCC. Ninth International Symposium on Computers and Communications*, vol. 1, pp. 244–251, 28 June-1 July 2004.

[10]   M. Buettner, G. V. Yee, E. Anderson, and R. Han, "X-MAC: A Short Preamble MAC Protocol for Duty-Cycled Wireless Sensor Networks," *ACM SenSys*, pp.307–320, November 2006.

[11]   A. Ker¨anen, J. Ott, and T. K¨arkk¨ainen, "The ONE Simulator for DTN Protocol Evaluation," *in Proc. of the 2nd International Conference on Simulation Tools and Techniques*, pp. 1–10, 2009.

[12]   W. Ye, J. Heidemann, and D. Estrin, "Medium Access Control With Coordinated Adaptive Sleeping for Wireless Sensor Networks," *IEEE/ACM Transactions on Networking*, vol. 12, no. 3, pp. 493–506, June 2004.

[13]   T. V. Dam and K. Langendoen, "An Adaptive Energy-Efficient MAC Protocol for Wireless Sensor Networks," *in Proc. of the 1st international conference on Embedded Networked Sensor Systems*, 2003.

[14]   P. Hurni and T. Braun, "An Energy-Efficient broadcast-ing scheme for unsynchronized wireless sensor MAC protocols," *in Proc. of the 7th International Conference on Wireless On-demand Network Systems and Services* (WONS), 2010, pp. 39–46.

[15]   F. Wang and J. Liu, "Duty-Cycle-Aware Broadcast in Wireless Sensor Networks," *in Proc. of IEEE INFOCOM*, pp. 468–476, April 2009.

[16]   Y. Gu and T. He, "Bounding Communication Delayin Energy Harvesting Sensor Networks," *in Proc. of IEEE 30th International Conference on DistributedComputing Systems*(ICDCS), pp. 837–847, 2010.

[17]   M. Miklo ′ s, K. Branislav, S. Gyula, and L. A ′ kos, " The Flooding Time Synchronization Protocol," *in Proc. of the 2nd International Conference on Embedded Networked Sensor Systems*, pp. 39–49, 2004.

[18]   S. Biswas and R. Morris, "ExOR: opportunistic multi-hop routing for wireless networks," *Computer Communication Review*, vol. 35, no. 4, 2005.

[19]   Q. Cao, T. Abdelzaher, T. He, and R. Kravets, "Cluster-Based Forwarding for Reliable End-to-End Delivery in Wireless Sensor Networks," *in Proc. of IEEE INFOCOM*, pp. 1928–1936, May 2007.

[20]   Y. Sun, O. Gurewitz, S. Du, L. Tang, and D. B. Johnson, "ADB: An Efficient Multihop Broadcast Protocol based on Asynchronous Duty-cycling in Wireless Sensor Networks," *in Proc. of the 7th ACM Conference on Embedded Networked Sensor Systems*, pp. 43–56, 2009.

[21]   S. Guo, Y. Gu, B. Jiang, and T. He, "Opportunistic Flooding in Low-duty-cycle Wireless Sensor Networks with Unreliable Links," *in Proc. of the 15th Annual International Conference on Mobile Computing and Networking*, pp. 133–144, 2009.

[22]   S. Lai and B. Ravindran, "On Multihop Broadcast over Adaptively Duty-Cycled Wireless Sensor Networks," *Distributed Computing in Sensor Systems*, pp. 158–171, 2010.

[23]   S. Lai and B. Ravindran, "Efficient Opportunistic Broadcasting over Duty-Cycled Wireless Sensor Networks," *in Proc. of IEEE INFOCOM*, 2010.

[24]   Y. Shao and J. Wu, "Understanding the Tolerance of Dynamic Networks: A Routing-Oriented Approach," *in Proc. of 28th International Conference on Distributed Computing Systems Workshops, ICDCS'08*, pp. 180–185, 2008.

[25]   H. Chen, L. Cui, and V. O.K.Li, "A Joint Design of Opportunistic Forwarding and Energy-Efficient MAC Protocol in Wireless Sensor Networks," *in Proc. of Global Telecommunications Conference, GLOBECOM*, 2009.

3

CHANNEL ASSIGNMENT ALGORITHMS FOR MRMC WIRELESS MESH NETWORKS

Mohammad A Hoque[1] and Xiaoyan Hong[2]

Department of Computer Science, The University of Alabama, USA
[1]mhoque@cs.ua.edu, [2]hxy@cs.ua.edu

ABSTRACT

The wireless mesh networksare considered as one of the vital elements in today's converged networks,providing high bandwidth and connectivity over large geographical areas. Mesh routers equipped with multiple radios can significantly overcome the capacity problem and increase the aggregate throughput of the network where single radio nodessuffer from performancedegradation. Moreover, the market availability of cheap radios or network interfaces also makes multi-radio solutions more feasible.A key issue in such networks is how to efficiently design a channel assignment scheme that utilizes the available channels as well as increases overall performance of the network. This paper provides an overall review on the issues pertaining to the channel assignment in WMNs and the most relevant approaches and solutions developed in the area. They include design challenges, goals and criteria; routing considerations, graph based solutions and challenges of partially overlapped channels. We conclude that the assignment of channels to the radio interfaces continuously poses significant challenges. Many research issues remain open for further investigation.

KEYWORDS

Channel assignment, wireless multi-hop routing, multiple radios and multiple channels, wireless mesh networks, partially overlapping channels.

1. INTRODUCTION

Wireless Mesh Network (WMN) provides a very reliable and cost efficient alternative for Internet connectivityover wide areasand enables ubiquitous computing environment through multi-hop relay[1]. In real world implementation of WMN, the total number of network interfacesis muchgreater than the number of frequency channels available for transmission. Moreover, each wireless node can have more than one interfaces or radios. This may lead to a topology where many mutually interfering links are assigned to the set of channels. This interference between concurrent transmissions can detrimentally degrade the throughput or performance of these networks. Therefore, as with cellular networks, the key factor for minimizing the effect of interference is the efficient reuse of radio frequency. One of the major issues concerned with WMN architecture supporting multiple radios and multiple channels (MRMC)is the channel assignment (CA) problem. Particularly for multi-hop networks, it is very complex to design an optimized CA algorithm that makes efficient utilization of available channels and at the same time minimizes the overall network interferences. In general, channel assignment algorithms shouldfacilitate multi-path routing amongwireless routers apart from minimizing interference onany given channel or from adjacent channels.

Existing channel assignment algorithms designedfor multi-radio multi-channel wireless mesh networks(MRMC-WMN) mostly deal with orthogonal or non-overlappingchannels. Recently the limited availabilityof orthogonal channelsin dense networks has motivated the wireless research community to studypartially overlapped channels (POC), which are consideredas a great potential for increasing the number ofsimultaneous transmissions and eventually upgrading thenetwork capacity; especially in the case of MRMC-WMN.

A substantial amount of research has been done so farwith multi-dimensional categories of channel assignmentschemes in wireless networks [2,3,9,12-20,39,40]. In this paper, variousaspects of channel assignment algorithm are discussedfrom different perspectives. Specifically, we outline objectives,design features that differ one from other solutionsand set them into different categories. We thendiscuss in details about a few representative solutions.Our emphasis then goes on graph based approaches andsolutions using POC. In all, our scope of this papergradually narrows down starting from the broad area ofchannel assignment (CA) to the depth of using POCs.

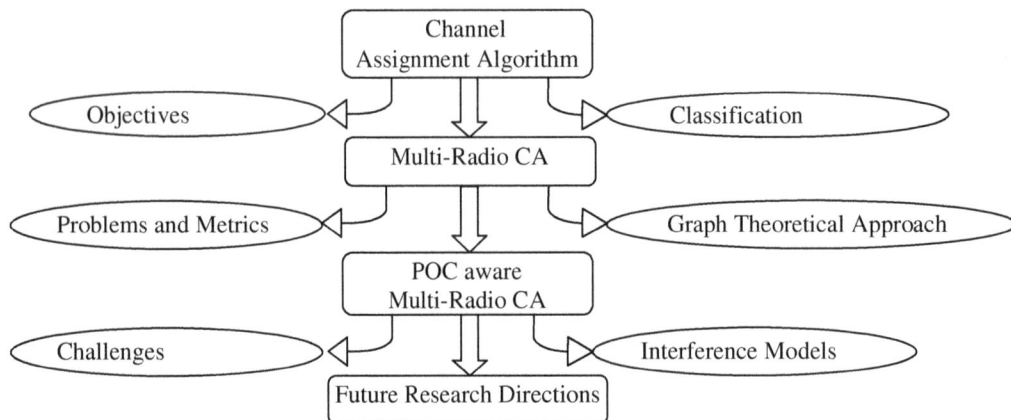

Figure 1.Organization of Channel Assignment Survey

The organization of the paper is illustrated in Figure1. The paper has three main parts. In the first part, consistingof Sections II and III, we discuss research issuesof channel assignment algorithms in general. In SectionII, we begin with outlining the common objectives inthe existing CA algorithms, and in Section III, we depictvarious ways that the classifications of the CA algorithmshave been used by many researchers. After that, in thesecond part, which consist of Sections IV, V and VI, weconcentrate on issues particularly related to the multipleradio environments. In section IV we argue in favor ofthe deployment of multi-radio communications followedby the choice of routing protocols associated with suchdeployment in section V. Later we present a graphtheoretical framework of formulating channel assignmentproblems in section VI. Then, in the third part, PartiallyOverlapped Channel (POC) related design issues areenumerated in section VII. Finally, we discuss the futureresearch directions in Section VIII and conclude thepaper in Section IX.

2. CHARACTERISTICSOF EFFICIENT CHANNEL ASSIGNMENT ALGORITHM

In the literature, solving channel assignment problems have been targeted to meet various design objectives. Some of these goals are described below.

One of the key objectives that need to be considered while designing a channel assignment scheme is to *minimize the network interference*. This interference minimization goal can either be implemented globally (in case of centralized schemes) or locally (in case of distributed schemes). It has been proved in the literature that the channel interference effects can cause a significant throughput loss in the network, especially if the design includes partially overlapped channels. Hence, most of the channel assignment algorithms should focus on this issue with severe importance.

All wireless networks are subject to capacity limitations due to many issues related to the characteristics of physical media. So a common goal for any wireless design is to focus on increasing capacity by applying innovative channel assignment schemes that can*maximize the overall network throughput*. Throughput is often regarded as the primary criterion to evaluate the efficiency of a new scheme. In fact, throughput is maximized by increasing the number of parallel transmission in a network. So, channel assignment algorithms should equally focus on throughput maximizing.

The IEEE 802.11 standard specifies multiple non-overlapping channels for use (3 in 802.11b/g, 12 in 802.11a). So the channel assignment scheme should aim into exploiting channel diversity to *maximize spectrum utilization*. Also, carefully allocating partially overlapped channels with proper interference model can further improve the channel utilization to maximum level. Therefore, researchers of wireless mesh networks are emphasizing on increasing channel diversity while designing channel allocation schemes.

Adaptation to changing traffic conditions is another important criterion for a well designed channel assignment scheme. An efficient channel assignment algorithm should not only maximize channel utilization but also *distribute the load equally* among different channels.

Inefficient channel assignment may lead to network partitions which ultimately deforms the original topology. So, *preserving the topology* by avoiding network partition is also an important goal for channel assignment algorithms.

3. CLASSIFICATION OF CHANNEL ASSIGNMENT SCHEMES

The channel assignment schemes can be classified based on different criteria and perspectives. Table I summarizes the classification followed by the description of each category thereafter. It is noteworthy that, these categories are not necessarily disjoint from each other. A particular type of scheme based on one criterion may fully or partially overlap with another type in different criteria.

Table 1. Classification of Channel Assignment Algorithms

Classification Criteria	**Types of Channel Assignment**

Channel Switching Frequency	a) Static/Fixed:
	  ▪ Common Channel Assignment (CCA)
	  ▪ Varying Channel Assignment (VCA)
	b) Dynamic
	c) Hybrid

Number of Radios	a) Single Radio b) Multiple Radio
Spectrum Utilization	a) Orthogonal Channels (OCs) b) Partially Overlapped Channels (POCs)
Topology Awareness	a) Centralized b) Distributed
Routing Dependency	a) Routing independent b) Routing dependent c) Joint Approach

Infrastructure	a) Access Point based
	b) Ad hoc based
	c) Hybrid approach
Granularity of Assignment	a) Per Packet Channel Assignment
	b) Per link Channel Assignment
	c) Per Flow Channel Assignment
	d) Per Component Channel Assignment

3.1 Based on Channel Switching Frequency

Skalliet. al. [40] proposed a taxonomical classification of various channel assignment schemes based on the criteria of channel switching frequency where the channel assignment schemes are divided into three main categories: fixed, dynamic and hybrid.

3.1.1 Fixed/Static Channel Assignment

Fixed or Static assignment schemes assign each radio to a channel for a relatively long period of time. The purpose of fixed channel assignment is to control the connectivity of the nodes.Das et. al [53] described some of the key issues related to static channel assignment algorithms. Fixed channel assignmentscheme has been further subcategorized into two types:

Common Channel Assignment (CCA) is the simplest among all the schemes where the network interfaces of each node are assigned to a common set of channels.   The primaryadvantage of this approach is that the network topology essentially remains identical tothatusing a single channel assignment scheme, while increasingthe network throughput by the use of multiple channels. However, in case where the number of orthogonal channels is greater than the number of radios in each node, the throughput gain may be limited andmay lead to inefficient channel utilization.

In case of ***Varying Channel Assignment (VCA)***,radios of different nodes are assigned to different sets of channels. However, assigning disjoint set of channels to the NICs may lead to network isolation and modified topology.An example of this type of algorithms is Connected Low Interference Channel Assignment (CLICA) [41].

3.1.2 Dynamic Channel Assignment

In Dynamic assignment schemes, any radiocan be assigned to any channel where theradios can frequently switch from one channel to another. The advantage of dynamic assignment is that it utilizes multiple channels with few interfaces. However, these approaches have the disadvantage of strict time synchronization requirement between the nodes. Other key challengesconstitute of channel switching delays and the need for signalling and coordination mechanisms for channel switching between a pair of nodes. These constraints impose practical challenges forimplementation in real networks.

3.1.3 Hybrid Channel Assignment

Hybrid channel assignment strategies combine both fixed and dynamic assignment strategies. Here, some radiosare assigned to a static channel whereas others can be dynamically switched between several channels.

3.2 Based on Number of Radios

When all the nodes in a WMN are equipped with ***single radio***, these channel assignment schemes are applicable. Advantages of this type are: (i) no complicacy of self-interference, (ii) channel selection algorithm is quite simple as only one channel has to be selected and finally

(iii) easy to implement. However, it also has drawbacks like: (i) less channel utilization (ii) no simultaneous transmission possible from a single node (iii) frequent channel switching

Currently thechannel assignment algorithms are targeted for meshnetworks with *multi-radio* environment. As multiplechannels are utilized at a time, channel utilization ismuch higher. Advantages of multi-radio scheme include(i) less channel switching and (ii) parallel transmissions.However, the channel selection algorithm is complex andinterference handling is also more difficult.

3.3 Based on SpectrumUtilization

Currently almost all channelassignment algorithms are designed with non-overlappingchannels or *Orthogonal channels*. This doesnot utilize all the available channel resources allocatedfor the specific IEEE 802 technology. For example,in case of IEEE 802.11 b/g/n, there are only 3 non-overlappingchannels out of 11 channels (in USA). Duringthe network overload period, there are not sufficientspectrum resources available when using only orthogonalchannels. This initiates the necessity of designing efficientschemes that can utilize all the available channelsin the spectrum.

Recently, a substantial amount of research is goingon with designing channel assignment algorithms with *Partially Overlapped Channels (POC)*. Some of the researchers alreadycame up with efficient algorithms that could handle theinterfering channels. But still questions exist about thefeasibility of implementing those schemes into currentindustry standard. We shall discuss the issues concerningthe POCs later.

3.4 Based on Topology Awareness

Centralizedchannel assignment algorithms have the global knowledgeabout the topology, either through global positioningsystem or though routing table information. Theyare mostly useful in case of infrastructure based wirelessnetworks like AP based networks. Centralized algorithmsare easy to implement, less overhead required for routingand node connectivity is determined by access points(APs). In other case, centralized channel assignment isalso applicable without APs when all the nodes have theglobal topology information. In most cases, centralizedalgorithms are either static or quasi-static.

Distributedchannel assignments are the ideal requirement for Adhoc networks. The distributed approach is more feasiblein realistic environments where the global informationfor centralized algorithm is not available. Our previous work [55] summarized a classification of MRMC channel assignment and routing algorithms on the basis of centralized and distributed categories.

3.5 Based on Routing Dependency

Most of the channel assignmentschemes are *independent of routing protocol*.These schemes work with any type of routing protocol,irrespective of proactive or reactive routing categories.Some channel assignmentschemes *depend on the type of routing protocol*. Thesealgorithms only work with the associated routing protocols.A recent trend is to design *jointrouting and channel assignment* schemes that optimizethe route by selecting the channels along the end to endpath. In such cases, channel information is also appendedin the routing table and broadcasted periodically. Inthese cross layer designs, efficient routing metric hasto be selected incorporating the channel interferencecharacteristics. An example of such joint approach is theKN-CA algorithm, by Xiaoguang Li et. Al. [24], whichis an enhancement of AODV protocol.

3.6 Based on Infrastructure

Channel assignment schemes that are particularly dependant on infrastructure or *based on access point* are mostly centralized.In that case the access point has the information of allthe nodes and their adjacent channels. In such case,the access point allocates the channel in a manner thatminimizes the overall interference and maximizes thethroughput and capacity.

On the other hand, *ad hoc mesh networks*lacks the information of global topology. Henceit is difficult to implement a centralized scheme with thelimited local information. Such centralized design basicallyimposes static channel assignment. Again, usingdistributed approach, the algorithm is prone to inaccuratetopological information which results into networkpartitioning.

In *hybridmesh networks*, nodes are connected in two ways, one isthe direct single hop connectivity with access point, andanother way is to route through other nodes to connectto a relatively less traffic loaded access point.This type of schemes is applied to areas where load density is high.

3.7 Based on Granularity

Per-packetchannel assignment requires more run-time control overheadfor scheduling each single packet with particularchannel. Hence, algorithms in this type are less efficientfor high loads. In [2], [6], Vaidyaet. al., described such a CA scheme where the radios switch from one channel to another in a small time scale. In reality, this type of scheme is not feasible for implementation because of the high overhead.

In*link-based channel assignment*scheme, channelis assigned to a link between a pair ofnodes, and all packets transmitted between these two nodes usethat particular channel for a certain period of time. Some of the algorithms of this type,focus on assigning channels by ensuringappropriate amount of bandwidth for each linkaccording to the expectedload. On the other hand, other schemesemphasize onminimizing link interference in the network. Several optimization models are also proposedin the literature for centralized channel assignment instatic WMNs, focusing on either maximizing the number simultaneously active links [56] or minimizing the overallinterferences among links.

In *flow-based channel assignment* scheme, a single channelis assigned to consecutive links along path from source to destination which defines a flow. As for example, So et al.[19]described a channel assignment schemethat binds separate channels to each of the flows ina single radio multichannel network.

Flowbased scheme is extended by Sivakumaret. al. in [20]to *component-based channel assignment*. A componentis formed by intersecting flows at a particular node and according to this approach an entire componentisassigned a single channel.

4. PROBLEMS WITH MULTI-RADIO CHANNEL ASSIGNMENT

The IEEE 802.11b/g/n standards provide 3 and 12 non-overlappingchannels that can be usedin parallel within a mesh network. If multiple radios can be installed on the same node to facilitate the simultaneous use of some of the channels, one can expect increased working bandwidth. Themarket availability of cheap NIC hardwarehas made the multi-radio solutions more feasible.Several research works [12,13,14] have proved that equipping a node with only 2 radios may increase the network capacity as well as throughput by a factor of 6 or 7. However, beside these benefits, there are a lot of problems associated withmulti-radio channel assignment. Throughout the followingsubsections, we address some of the critical problemsrelated to MRMC design.

4.1 Interference Minimization

Although multi-radio wireless nodes cansignificantly uplift the performance of WMN,there is a critical trade-off to be made between maximizingconnectivity and minimizing interference. Thekey factors to consider are the co-channeland adjacent channel interference due tothe close proximity of the radios equipped on a single node, and those due to the transmissions from neighbouring nodes [35]. The co-channel interference prohibits a particular channel to be used more than once by two links within the interference range simultaneously. The adjacent channel interference determines the total number of usable channels within the neighbourhood (defined by the transmission range). In order to minimize thenetwork interference, a suitable interference model hasto be designed in accordance with the assignable channelsuper set. For example, an interference model which iscapable of handling the self interferenceproblem maynot be suitable for POC based design.

4.2 Channel Switching Delay

One of the key challenges in multi-radio environmentinvolve channel switching delay which is typically inthe order of several milliseconds. This mandates tight coordination mechanisms forchannel switching between nodes. Hence, the frequencyof channel switching greatly impacts the efficiency andthroughput of the network.

4.3Interdependency with Routing Protocol

As a matter of fact, routing and channel assignment are interdependent. A routingprotocol selects a path from the source to the destination,and forwards traffic to each link along the path,while channel assignment determines the individual channel that each linkshould use. In other words, CA determines the connectivitybetween two nodes as two radios can only communicatewhen they are tuned to a common channel.Hence channel assignmentultimately determines the network topology. Again, as weknow, routing decisions are dependent on the networktopology which implies thatchannel assignment has a direct impact on routing.Experiments have shown that, dynamically adjusting the channelaccording to thetraffic status can achieve better result, which again proves that routing and channel assignment are tightly coupled.

4.4Issues with Joint Channel Assignment and Routing

In order to maximize the performance gain in MRMC-WMN, joint implementation of routing and channel assignment is very important. Traditional wireless routing protocols [7,8,11] may not provide optimized performance without incorporating integration with CA. Wireless researchers focussing on cross layer protocol design mostly deal with integrating routing with CA. Some of these schemes are designed as centralised algorithm [14, 24, 27, 41] while others considered distributed mode [9, 37]. However, there are several challengesin effectively designing algorithms for joint CA androuting, especially in a distributed fashion. More complicacyarises when the network is a heterogeneous typeof multi-radio wireless networks. Below we mentionsome of the critical issues while designing a joint CAand routing algorithm:

For any routing protocol whether or not integrated with CA, a*routing metric* needsto be concretely defined as a quantitative measurement of the performance gain.In case of joint CA and routing, most of these metrics are defined as compound metric derived from other elementary routing metrics. One such algorithm ofthis type [37] defines a metric named Channel Cost Metric(CCM) thatcomputes the expected transmission costweighted by channel utilization.CCM quantifies the effect of channel interferences along withthe benefit of channel diversity.

Another major issue arises in networks with **heterogeneous radios**operating with different transmission power and frequency.It can be possible thatthere be no common radio or commonchannel supported in the whole network for both datatransmission and signalling (e.g., routing message), leading to network partitioning.Bhandari and Vaidya [38,42] revealed many issues particularly applicable for networks with heterogeneous radios. Further, **reducing the protocol overhead** for a distributed algorithm in such a heterogeneous wirelessenvironment presents significant challenges for thejoint implementationof CA and routing.

5. CHOICE OF ROUTING METRIC INTEGRATED WITH CA

5.1 Evolution of Routing Metrics

In this section, we discuss the routing metrics that havebeen widely accepted for mesh networks in a hierarchical representationbased on their derivation. Some of the well knownrouting metrics are: hop count, RTT, ETX[4], ETT[5], WCETT[5],EDR [10], CCM[37], MCR[15], MIC[18], ILA[48] and iAWARE[50]. Addagaddaet. al. [47] summarized some of the notable features of these routing metrics and proposed modifications over ILA and iAWARE. All these metrics aretopology-dependent and mostmetricswereproposed as improvement over some other previous metrics. Figure 2shows a hierarchical representation of the metrics based on their derivation.

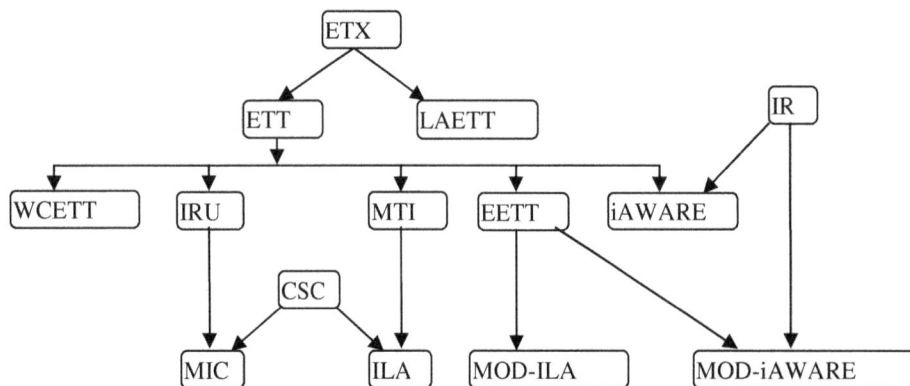

Figure 2. The hierarchical representation of the metric

We also tabulated some of the interesting characteristicsof these metrics under Table 2. These characteristicsgave us a foundation to classify the metrics from twodifferent perspectives, i.e. we categorized the routingmetrics based on isotonicity and also based on interferenceconsideration.

Table 2.Summary of characteristics of the routing metrics used in wireless networks

Characteristics	Hop	RTT	ETX[4]	ETT[5]	WCETT[5]	CCM[37]	MIC[18]	EETT[21]	LAETT[49]	ILA[48]	iAWARE[50]	Mod-iAWARE[47]
Multi-channel Support	X	X	X	X	Y	Y	Y	Y	Y	Y	Y	Y
Intra-Flow Interference	X	X	Y	X	Y	Y	Y	Y	X	Y	Y	Y
Inter-flow Interference	X	X	X	X	X	Y	Y	Y	X	Y	Y	Y
Load balancing	X	Y	X	X	X	Y	X	X	Y	X	X	Y
Link loss ratio	X	X	Y	Y	Y	Y	Y	Y	Y	X	Y	Y
Throughput	X	X	Y	Y	Y	Y	Y	Y	Y	Y	Y	Y
Transmission Rate	X	X	X	Y	Y	Y	Y	Y	Y	Y	Y	Y
Link Capacity	X	X	X	Y	Y	Y	Y	Y	Y	Y	Y	Y
Multi-Radio Support	X	X	Y	Y	Y	Y	Y	Y	Y	Y	Y	Y
Heterogenous Radio	X	X	X	X	X	Y	X	X	X	X	X	X
Agility	Y	Y	X	X	X	X	X	X	X	X	X	X
Isotonicity	Y	Y	Y	Y	X	X	Y	Y	Y	X	X	Y

5.2 Classification Based on Isotonicity

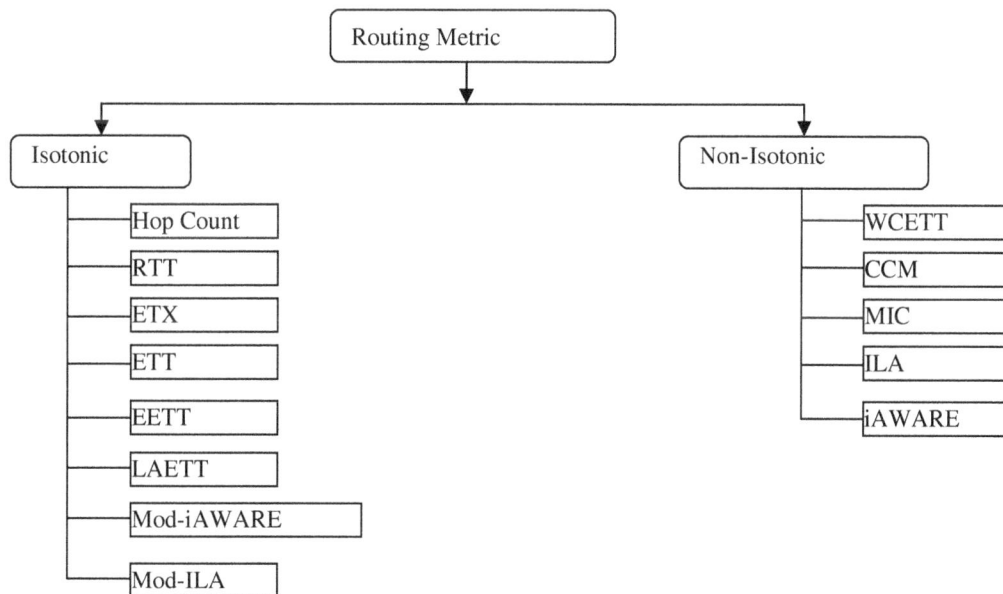

Figure 3. Classification based on Isotonicity

In order to calculate the minimum cost path, most routing protocols follow certain variations of efficient algorithms, like Bellman- Ford or Dijkstra's algorithms. Even if a metric guarantees that its minimum costroute has good performance, there is no assurance ofhaving an efficient algorithm to compute the path cost based on the metric. The property that ensures the existence of such efficient algorithmis called isotonicity [45]. Based on this property, routing metrics can broadly be categorized into two classes, namely i) Isotonic and ii) Non-Isotonic. Figure 3 shows the classification of some of the common routing metrics on the basis of isotonicity.

5.3 Classification Based on Interference

While designing a routing metric, two types of interferences are needed to be considered in a mesh network:

Intra-flow interference occurs while the network interfaces of two or more consecutive links belonging to a single path or flow operate on the same channel.This type of interferences can be mitigated by applying channel diversity; for example, by selecting non-overlapping or orthogonal channels for subsequentlinks. Typically the interference range isgreater than transmission range beyond immediate neighbors. This might result into interference among non-adjacent links operating on same channel in a multi-hop path.

Inter-flow interference is caused by interference generated from other flows that are operating on the same channels. Due to the involvement of multiple flows and routes, controlling inter-flow interference is more complicated than intra-flow interference. Based on the consideration of these interferences, routing metrics can be classified to four categories as shown in Figure 4.

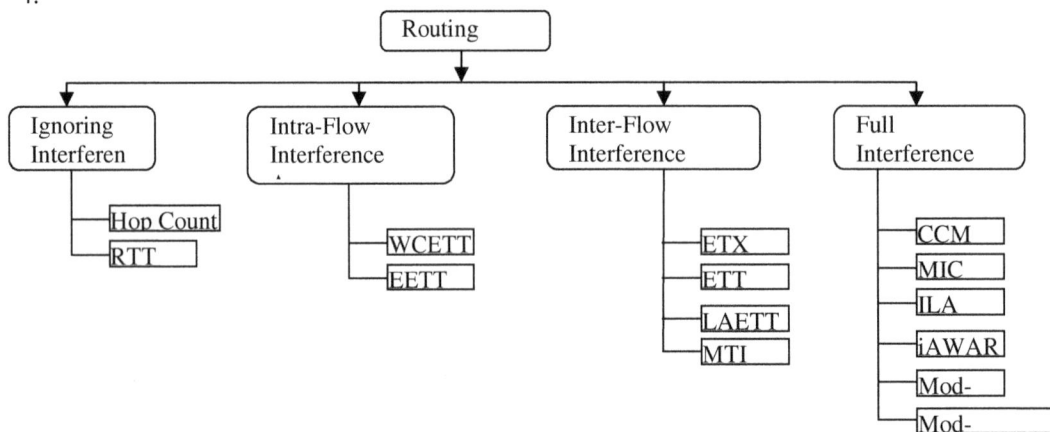

Figure 4. Classification based on Interference

6. GRAPH THEORETICAL FRAMEWORK FOR CHANNEL ASSIGNMENT

Graph based algorithms have been widely used in many channel assignment algorithms, irrespective of number of radios and channels. The network topology input is generally specified as a connectivity graph. The connectivity graph may be simple undirected graph or multi-graph depending on the number of radios and link topology. This connectivity graph can be converted into an intermediate graph, which generally takes the form of a conflict graph, characterizing the impact of mutual link interferences. For example, when coloring algorithms are used, this conflict graph is fed as input to the graph coloring algorithm which ultimately finds the channel mapping solution for the links. The method is depicted in Figure 5.

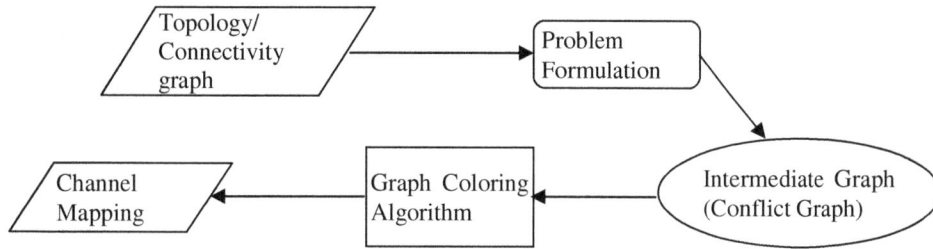

Figure 5. Framework for channel assignment

6.1 Graphical Representation of Channel Assignment Problems

Researchers have developed many approaches to design solutions for channel assignment. To formulate the channel assignment problems, different versions of conflict graphs are commonly used to characterize the interference constraints, whereas the application of various graph coloring algorithms has become a popular practice in selecting channels. Below we mentioned some of the graphical models that are very widely used during problem formulation of multi-radio channel assignment:

6.1.1 Simple Conflict Graph

A simple conflict graph $G_c(V_c,E_c)$ is a graph derived from the original network topology graph where each vertex V_c represents a communication link or egdeof the topology. There is an edge between two vertices of conflict graph only if the corresponding links in the topology are mutually interfering. An illustration is given in Figure 6. Figure 6(a) shows the original network topology where the three links ij, jp and pqare represented as vertices in Figure 6(b). Here, all the three links interfere with each other because of the close proximity and hence all the three vertices in conflict graph are connected.

Figure 6(a). A four node network

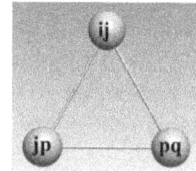

Figure 6(b). corresponding conflict graph

6.1.2 Weighted Conflict graph

Some researchers represent the interference effect through assigning various weights to the edges of conflict graph. These types of graphs are known as Weighted Conflict graphs. These weights are assigned based on the extent of interference calculated from appropriate interference model. Two well known algorithms, CLICA [41] and CoSAP [30] are formulated using these models. Of them, the latter is applicable cognitive radio networks.

6.1.3 Multi-radio Conflict graph

 K. N. Ramachandran et al [12], introduced the notion of Multi-radio Conflict graph (MCG). The authors extended the simple conflict graph to model multi-radio mesh routers (Figure 7). In this model,edges between individual radiosare represented as vertices instead of representing edges between the nodes..Figure 7(a) shows a wireless network with four nodes A,B,C and D where node C is equipped with 2 radios while the rest have single radio. Figure

7(b) is the corresponding simple conflict graph while Figure 7(c) shows the multi-radio conflict graph. In the multi-radio conflict graph, all the links connected to node C are represented with two edges, each for an individual radio.

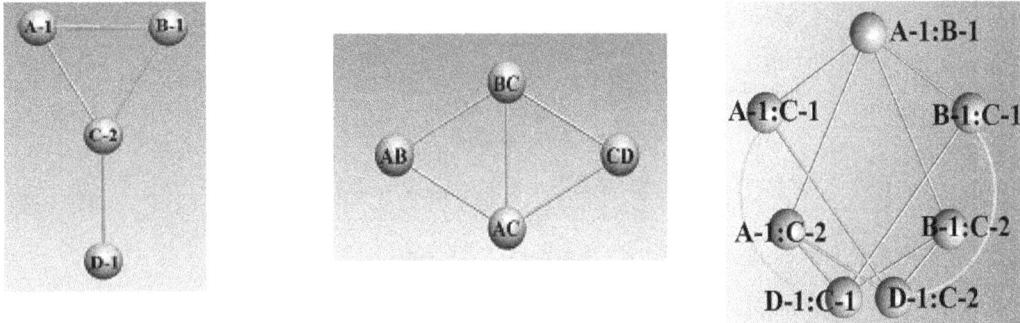

Figure 7(a). A simple network topology

Figure 7(b). Corresponding conflict graph

Figure 7(c). Corresponding multi-radio conflict graph

6.1.4 Resource Contention Graph(RCG)

W. Wang et al. [16] proposed the notion of Resource Contention Graph (RCG)which captures various contention regions in the network topology by identifying all the maximum cliques in the interference graph. The authors described a framework thatrepresents the capacity of a multichannel network when the topology is known. The framework is formulatedas an ILP problem wherethe solution of the problemdetermines the maximum possible spectrumusage for a given topology underchannel and radio constraints. For any specific traffic pattern, the framework provides an upper bound on throughput with optimal routingdecisions. Initially the resource contention graph is generated from thetopology graph. Then a max-flow-likegraph is constructed using the resource contention graph. The Max-flow graph is an extended version of the RCG which is generated by adding a source vertex s and a sink vertex t.For example, Figure8(a) is a topology consisting of 4 nodes. Figure8(b) illustrates the corresponding network flow model. The edge capacity for the first three levels is N, which is the number of channels and the edges of the last two levels have a capacity of K, which is the number of radios.

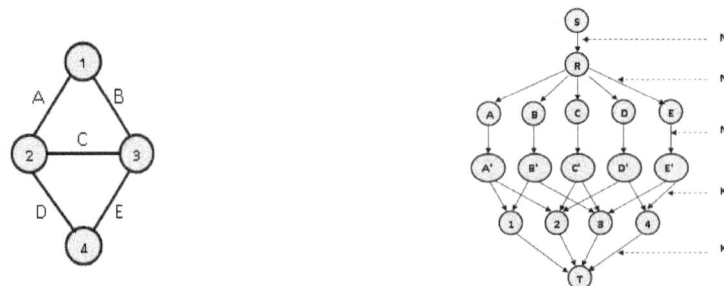

Figure8(a).A topology of 4 nodes Figure8(b).Resource Contention Graph

6.1.5 Layered Graph Model

C. Xin et al.[17] proposed a layered graph model to jointly optimize routing and channel assignment. In their model, each layer corresponds toa particular channel. The entire topology is represented using multiple layers of nodes where the number of layer is equal to total number of channels. A single network node is shown as a collection of virtual nodes residing in each layer. Vertical edges between layers connect the virtual nodes. The weights of the virtual edges are typicallysetwith a low cost which makes the routing protocol prefer a path with dynamic channel switching. Practically, the cost of the vertical edges should be equal to the cost of channel switching delay. The horizontal edges that belong to the same layer (channel) are the actual cost of air propagation delay. Figure 9 illustrates a simplified layered model of three channels withfour wireless nodes A, B and C, in which A and Dare a communicating pair. The routing path switches from channel 1 to channel 3 at node B and again switches from channel 3 to channel 2 at node C.

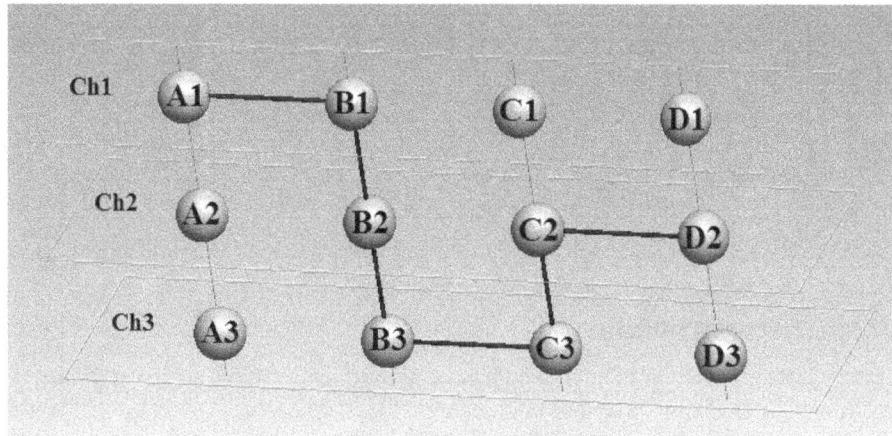

Figure9. Layered graph Model

6.2 Coloring Algorithms

Utilizing the different forms of conflict graphs described in the previous section, colors (i.e. channels) have to be assigned to the vertices of the conflict graph (which correspond to the links in the connectivity graph) so that an objective function is optimized. Typical objective functions range from minimizing the difference between the largest and the lowest used colors while avoiding interference to minimizing interferences using a given number of colors. For arbitrary networks, the resulting vertex coloring problems are computationally intractable (i.e., NP-hard). Therefore, the channel assignment problem is usually addressed by means of heuristic approaches, like genetic algorithms, taboo search, saturation degree, simulated annealing etc. Some researchers [52] tend to use polynomial time approximation schemes in greedy approach. Some of the common coloring or partitioning algorithms used to solve the channel assignment problems are Max K-Cut algorithm [32], MIN-MAX k-PARTITION [53], Distance-2 Edge Coloring/Strong Edge Coloring [43] etc.

7. POC AWARE MULTI-RADIO CA DESIGN CHALLENGES

7.1 Improving spectrum utilization through partially overlapped channels (POCs)

Existing channel assignment algorithms designed for multi-radio multi-channel wireless mesh networks (MRMC-WMN) mainly deal with orthogonal or non-overlapped channels. In fact, due to the adverse effects of adjacent channel interference, almost all channel assignment algorithms use orthogonal channels alone. In reality, the smallnumber of orthogonal channelsposes major challengesin dense networks. For example, the 802.11b/g standards define a total of 11 channels in the US, out of which only 3 areorthogonal (Figure 10). Most residential usersand WLAN administrators tune their network interfaces to one of these 3 channels only. Thus two potentially interfering nodes can be assignedto the same channel.This ultimately leads to a wastage of wireless spectrum capacity. Recently it has been revealed that using partially overlapped channels can lead to better utilization of the spectrum. However, an ad-hoc use of POCs can actually degrade performance. Many recent works[23,28,33,34,56, 60,61,62,63,64] have shown that the partially overlapped channels (POC) have a great potential for increasing the number of simultaneous transmissions for MRMC-WMN.

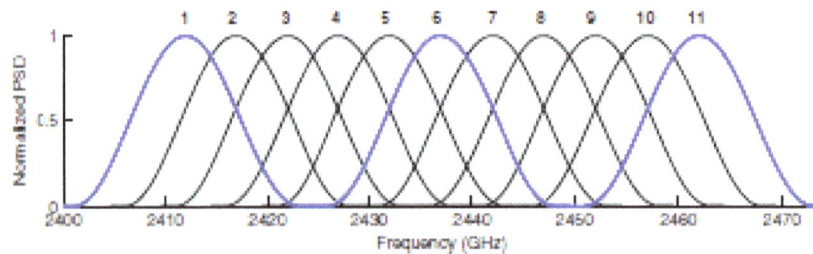

Figure 10. IEEE 802.11b/g channel distribution showing the 3 orthogonal channels in blue [65]

7.2 Self Interference Problem

One most critical challenge in POC based channel assignment is to overcome *self-interference problem*. Links connected to a single node cannot be assigned to channels with overlapping frequency bandwidth due to this problem. To the best of our knowledge, only one of the schemes [23] has identified this problem associated with multi-radio channel assignment. Due to this problem the maximum number of parallel transmission from a single node must be restricted to the number of maximum orthogonal channels available, which is 3 in case of IEEE 802.11b/g. Researchers need to concentrate on the severity of this problem and eventually negotiate with this self- interference issue in addition to dealing with Co-Channel interference and ACI. Otherwise, most algorithms would be overestimating the goodness of respective results. For this reason, choosing the appropriate interference model is also very important.

7.3 Choice of Interference Model

A fundamental difference between a wireless network and its wired counterpart is that wireless links may interfere with each other, resulting in performance degradation. As a result, there have been numerous researches on wireless networks considering interference between wireless links. Out of several kinds of interference, handling co-channel interference is relatively simpler because many of the wireless link layer protocols use contention resolution mechanisms like RTS-CTS which easily detects if the transmitting channel is busy or not. On the contrary, adjacent channel interferences(ACI) are difficult to detect using channel contention mechanisms because in most cases these ACIs contribute as background noise and reduce the signal to noise ratio. Below, we mention some of the possible choice and alternatives while considering the interference model in a POC based channel assignment algorithm:

Among all the interference models, the *Binary Interference Model* is the simplest one. The model defines that two links can be either interfering or non-interfering without quantifying the extent of interference among each other. Hence, this phenomenon can be represented as a binary condition. But researchers have proved that this 0/1 assumption in case of interference is not true for most cases. The network throughput depends on the actual amount of frequency overlap, distance between nodes and signal to noise ratio which quantifies the interference. Therefore, this model is not meaningful while considering the case of POC based design.

Gupta et. al [58] proposed two important interference model that has been widely used in wireless communication and information theory. These two models, namely protocol model and physical model, have been studied extensively in the literature by subsequent researchers. In*protocol model*,a geographical boundary or interference range is defined for each receiver within which a receiver may perceive interference from other potential transmitters residing inside the boundary where the interfering transmissions are also on the same channel. Hence, this model can capture the effect of co-channel interference but not ACI. On the other hand,in *physical model*, the interference is mathematically calculated from the signal to noise ratio. Here, a transmission is considered successful when the signal to noise ratioperceived by the receiver exceeds a minimum threshold value after accumulating noise signals contributed byall other transmitters. In this model,the choice of threshold is an important tunable parameter for actual interference measurement. Comparing, protocol and physical model, the latter is obviously the more accurate but the computational complexity is too high. On the other hand, protocol model is easy to calculate but may lead to erroneous results due to inability to capture ACI effect.

A *channel interference cost function*, proposed by Ko et al. [25], provides a measure of the spectral overlapping level between two partially overlapped channels. The interference cost between channel a and b, denoted by $f(a,b)$, is defined as$f(a,b) \geq 0$ and $f(a,b) = f(b,a)$, where a value of 0 indicates that channels are non-interfering.The value of $f(a,b)$ decreases as the frequency separation between the two channelsincrease. An example of a simple cost function defined using a single tunable parameter$\delta$is: $f(a,b) = \max(0, \delta - |a - b|)$ where δcan be defined as the minimum channel separation between two non-overlapping channels. For IEEE 802.11b/g, $\delta = 5$. For example, if $a=7$ and $b=4$, then $f(7,4) = \max(0, 5 - 3)= 2$. Again, for $a=9$ and $b=2$, $f(9,2)= \max(0, 5 - 3)$, which means no interference at all. Due to the simplicity of this cost function, it is also easy to implement in a channel assignment algorithm as a measure of partial interference.

Interference-factor or simply *I-factor*,proposed by the current authors [57], can also be regarded a suitable measure of the extent of channel overlap. If $P_i$denotes the power received at a given location of a particular signal on channel i, and $P_j$denote the power received of the same signal at the same location on channel j, then the interference factor between i and j,$I(i,j)$ is defined as $\frac{P_i}{P_j}$. $I(i,j)$ gives the fraction of a signal's power on channel j that will be received on channel i. I-factor can be calculated analytically as well as empirically and does not depend on the radio propagation properties of the environment (i.e. open space or indoors). It depends on the extent of frequency overlap between the signals on channels i and j. Hence, this is a suitable choice for POC based interference models. For example, some POC based channel assignment schemes have been proposed [56, 57, 63] using this interference model. Table III below shows I(i; 6) normalized to a scale of 0 . . . 1.

Table 3.*I (i,6)* Values [56]

Channel	1	2	3	4	5	6	7	8	9	10	11
Normalized SNR (I-factor)	0	0.22	0.60	0.72	0.77	1.0	0.96	0.77	0.66	0.39	0

An innovative *Channel Overlapping Matrix Model* has been introduced by A. Hamed Rad et. al [28]. The model captures the interference using a channel overlapping matrix. For example, let us consider an MRMC-WMN where N denotes the set of wireless routers whereeach router is equipped with I NICs. There are a total of C channels available for transmission. For any two routers $a, b \in N$, a $C \times 1$ channel assignment vector is defined $\overline{x_{ab}}$. If router a, communicates with router b over the ith channel, then the ith element in $\overline{x_{ab}}$ is equal to 1; otherwise, it is equal to zero. As for example, a router a islinked with router b through the 2nd channel where C = 5. This implies, $\overline{x_{ab}} = [0\ 1\ 0\ 0\ 0]^T$. Let, m and n are two of the available channels within the frequency band. To mathematically model the overlapping effect among different channels, the authors defined a symmetric $C \times C$ channel overlapping matrix W. The entry in the m-th row and the n-th column of W is denoted by scalar w_{mn} and is defined to be as follows:

$$w_{mn} = \frac{\int_{-\infty}^{\infty} F_m(\omega)F_n(\omega)d\omega}{\int_{-\infty}^{\infty} F_m^2(\omega)\,d\omega}$$

Where $F_m(\omega)$ and $F_n(\omega)$ denote the respective power spectral densities on channels m and n.

8. FUTURE RESEARCH DIRECTIONS

8.1 Multi-Radio Multi-Channel Concerns

Despite significant amount of research [16], the **network capacity** of WMNs is still a challenging topic. Although Vaidyaet. al. [22, 43, 44] characterized network capacity in terms of number of channels and radios as well as switching delay, more conditions can be added such as heterogeneous radios, mobility of nodes. On the other hand, Wang et. al. [16] proposed a framework to maximize overall capacity based on graph theoretical approach.In addition, as of to-date, no MRMC protocol exploits the **multi-rate capability** of current 802.11 wireless cards. By considering only homogeneous links, the problem becomes much simpler. However, a protocol with adaptive rates can achieve better performance.

The **channel switching delay** is an important concern for channel assignment schemes that switch the radio interfaces very frequently. Despite of significant improvement in wireless networking hardware, channel switching delay is still in the order of millisecond which is considered as an overhead for overall end-to-end delay. On the other hand, using a static channel assignment approach to avail the benefits of reduced overhead and stable topology will lack from the capacity improvement gained by MRMC environment. Therefore, a well estimated tradeoff is necessary to overcome the problem arising from switching overhead.

8.2 POC Aware Design

The wireless literature still lacks an efficient**POC based dynamic and distributed algorithm**, a algorithm that can handle channel switching for each node. Though some static schemes have been designed with POC [56, 60, 61, 62, 63, 64], more emphasis should be on dynamic versions.**No existing simulator** is capable of simulating such MRMC networks that involve interference calculated from POCs. Hence current popular simulators might be extended with features supporting POC channel model and network protocolsdesigned for partially overlapped channels. As of this date, there is no **joint routing and channel assignment** algorithm designed with POCs. Polynomial time approximation schemes are often considered as feasible solution in this area where many critical factors, such as compound routing metrics that characterizes appropriate interference model, should be handled.

Further, tuning the interference tolerance level by carefully *adjusting the SINR threshold* value is of great importance. A higher threshold value will definitely give better transmission quality with low interferences and noises, but will have higher probability of retransmission and low throughput, whereas, a small threshold will generate higher interference and degrade the quality of signal reception. Thus a tradeoff has to be made in case of deciding the SINR threshold value.

9. CONCLUSION

In this paper we have identified the key challenges and research approaches associated with assigning channels to radio interfaces in multi-radio wireless mesh networks. We have provided the goals and objectives of an efficient algorithm, classification of existing schemes and comparative analysis of different schemes. We presented the challenges involved with multi-radio and POC based design. In the end, we outlined important open research issues for future investigations.

REFERENCES

[1] E. Hossain, Kin KwongLeung (2008) Wireless Mesh Networks: Architectures And Protocols, Springer

[2] P. Kyasanur and N.H. Vaidya (2006) Routing and Link-layer Protocols for Multi-Channel Multi-Interface Ad Hoc Wireless Networks. In MobiCom.

[3] J.Crichigno,M. Wu, W.Shu (2008) Protocols and architectures for channel assignment inwireless mesh networks, Ad hoc Networks, Vol-6.

[4] D. S. J. D. Couto, D. Aguayo, J. Bicket, and R. Morris (2003) A high-throughput path metric for multi-hop wireless routing. In MobiCom.

[5] R.Draves, J. Padhye, and B. Zill (2004) Routing in multi-radio, multihop wireless mesh networks. In MobiCom.

[6] J. So, N.H. Vaidya (2004) Multi-Channel MAC for Ad Hoc Networks: Handling Multi-Channel Hidden Terminals Using A Single Transceiver. In MobiHoc.

[7] P. Jacquet, P. Muhletaler, T. Clausen, A. Laouiti, A. Qayyum, and L. Viennot (2001) Optimized Link State Routing Protocol for Ad Hoc Networks. In IEEE INMIC.

[8] D. B. Johnson and D. A. Maltz (1996) Dynamic source routing in ad hoc wireless networks. In Mobile Computing, Vol-353.

[9] X. Lin and S. Rasool (2007) A distributed joint channel-assignment, scheduling and routing algorithm for multi-channel ad hoc wireless networks. In IEEE INFOCOM'07.

[10] J. C. Park and S. K. Kasera (2005) Expected Data Rate: An Accurate High-Throughput Path Metric For Multi-Hop Wireless Routing. In IEEE SECON.

[11] C. E. Perkins, E. M. Royer, and S. Das (2003) Ad-hoc On Demand Distance Vector (AODV) Routing. http://www.ietf.org/rfc/ rfc3561.txt.

[12] K. N. Ramachandran, E. M. Belding, K. C. Almeroth, and M. M. Buddhikot (2006) Interference-aware channel assignment in multi-radio wireless mesh networks. In Infocom.

[13] A. Raniwala and T. Chiueh (2005) Architecture and algorithms for an ieee 802.11-based multi-channel wireless mesh network. In IEEE Infocom.

[14] A. Raniwala, K. Gopalan, and T. ckerChiueh (2004) Centralized channel assignment and routing algorithms for multi-channel wireless mesh networks. SIGMOBILE Mob.Comput.Commun.Rev. Vol-8.

[15]PradeepKyasanur and NitinVaidya (2005) "Multi-Channel Wireless Networks: Capacity and Protocols," Technical Report, University of Illinois at Urbana-Champaign.

[16] W. Wang and X. Liu (2006) A Framework for Maximum Capacity in Multi-channel Multi-radio Wireless Networks.In IEEE CCNC.

[17] C. Xin and C.-C.Shen (2005)A Novel Layered Graph Model for Topology Formation and Routing in Dynamic Spectrum Access Networks. In IEEE DySPAN.

[18] Y. Yang, J. Wang, and R. Kravets (2005) Interference-aware Load Balancing for Multihop Wireless Networks. In Tech Report UIUCDCS-R-2005-2526, Department of Computer Science, UIUC.

[19] J. So, N.H. Vaidya (2004) A Routing Protocol for Utilizing Multiple channels in Multi-hop Wireless Networks with a Single Transceiver, UIUC Technical Report.

[20] R. Vedantham, S. Kakumanu, S. Lakshmanan and R. Sivakumar (2006) "Component Based Channel Assignment in Single Radio, Multi-channel Ad Hoc Networks," InMobiCom.

[21]Weirong Jiang, ShupingLiu, Yun Zhu and Zhiming Zhang (2007) Optimizing Routing Metrics for Large-Scale Multi-Radio Mesh Network, In WiCom.

[22] P. Kyasanur, N. Vaidya (2005) Capacity of multi-channel wireless networks: Impact of number of channels and interfaces, In MobiCom.

[23] ZhenhuaFeng, Yaling Yang (2008) How Much Improvement Can We Get From Partially Overlapped Channels? , In IEEE WCNC.

[24] Xiaoguang Li, ChangqiaoXu (2009) Joint Channel Assignment and Routing in Real Time Wireless Mesh Network, In IEEE WCNC.

[25] Bong-Jun Ko Vishal, MisraJitendra, Padhye Dan Rubenstein (2007)Distributed Channel Assignment in Multi-Radio 802.11 Mesh Networks, In IEEE WCNC.

[26] Ka-Hung Hui, Wing-Cheong Lau, On-ChingYue (2007) Characterizing and Exploiting Partial Interference in Wireless Mesh Networks, In IEEE ICC.

[27] MansoorAlicherry,Randeep Bhatia, Li (Erran) Li (2005) Joint Channel Assignment and Routing for Throughput Optimization in Multiradio Wireless Mesh Networks, In MobiCom.

[28] A. HamedMohsenian Rad, Vincent W.S Wong (2007) Partially Overlapped Channel Assignment for Multi-Channel Wireless Mesh Networks, In IEEE ICC.

[29] Hua Yu, PrasantMohapatra, Xin Liu (2008) Channel assignment and link scheduling in multi-radio multi-channel wireless mesh networks, ACM Mobile Networks and Applications, Vol. 13.

[30] Anthony Plummer Jr., TaoWu, SubirBiswas (2007) A Cognitive Spectrum Assignment Protocol using Distributed Conflict Graph Construction, In IEEE MILCOM.

[31] Seongho CHO, Chong-kwon KIM (2008) Interference-Aware Multi- Channel Assignment in Multi-Radio Wireless Mesh Networks, IEICE Transactions on Communications Vol. E91-B(5).

[32] AnandPrabhu Subramanian, Himanshu Gupta, Samir R. Das (2008) Minimum Interference Channel Assignment in Multi-Radio Wireless Mesh Networks, IEEE Transactions on Mobile Computing, Vol. 7.

[33] Arunesh Mishra, Eric Rozner, Suman Banerjee, William Arbaugh (2005) Exploiting Partially Overlapping Channels inWireless Networks: Turning a Peril into an Advantage, Internet Measurement Conference.

[34] Arunesh Mishra, VivekShrivastava, Suman Banerjee, William Arbaugh (2006) Partially Overlapped Channels Not Considered Harmful, ACM SIGMETRICS Vol. 34.

[35] Vijay Raman, Nitin H. Vaidya (2009) Adjacent Channel Interference Reduction in Multichannel Wireless Networks Using Intelligent Channel Allocation, Technical Report, ICWS, UIUC.

[36] Stefano Avallone, Ian F. Akyildiz (2008) A channel assignment algorithm for multi-radio wireless mesh networks, Computer Communications, Vol. 31.

[37] Haitao Wu et. al (2006) Distributed Channel Assignment and Routing in Multiradio Multichannel Multihop Wireless Networks, IEEE Journal on Selected Areas in Communications, Vol. 24.

[38] VartikaBhandaria, NitinVaidya (2008) Heterogeneous Multi-Channel Wireless Networks: Routing and Link Layer Protocols, Mobile Computing and Communications Review, Vol.12.

[39] Ian F. Akyildiz, X. Wang (2005) A Survey on Wireless Mesh Networks, IEEE Radio Communications.

[40] HabibaSkalli, SamikGhosh and Sajal K. Das, Luciano Lenzini, Marco Conti (2007) Channel Assignment Strategies for Multiradio Wireless Mesh Networks: Issues and Solutions, IEEE Communications Magazine.

[41] M. Marina and S. R. Das (2005) "A Topology Control Approach for Utilizing Multiple Channels in Multi-Radio Wireless Mesh Networks," In Proc. Broadnets.

[42] V. Bhandari and N. H. Vaidya (2007) Heterogeneous multi-channel wireless networks: Scheduling and routing issues. Technical Report, UIUC.

[43] V. Bhandari and N. H. Vaidya (2007) Capacity of multichannel wireless networks with random (c, f) assignment.In MobiHoc.

[44] V. Bhandari and N. H. Vaidya (2007) Connectivity and Capacity of Multichannel Wireless Networks with Channel Switching Constraints.In IEEE INFOCOM.

[45] Y. Yang, J. Wang, R. Kravets (2005) Designing routing metrics for mesh networks, In IEEE WiMesh.

[46] C.L. Barrett, G.Istrate, V. S. A.Kumar, M.V. Marathe, S.Thite, S.Thulasidasan (2006) Strong Edge Coloring for Channel Assignment in Wireless Radio Networks, In IEEEPerCom.

[47] Bharat Kumar Addagada,VineethKisara and Kiran Desai (2009)A Survey: Routing Metrics for Wireless Mesh Networks.

[48] DevuManikantanShila and Tricha Anjali (2008) Load-aware Traffic Engineering for Mesh Networks, Computer Communications, Vol-31.

[49] HervéAïache, Laure Lebrun, Vania Conan and Stéphane Rousseau (2008) LAETT A load dependent metricfor balancing Internet tra-c in Wireless Mesh Networks, In IEEE MASS.

[50] AnandPrabhuSubramanian ,Milind M. Buddhikot and Scott Miller (2006) Interference Aware Routing in Multi-Radio Wireless Mesh Networks.In IEEEWiMesh.

[51] Weirong Jiang, Shuping Liu, Yun Zhu and Zhiming Zhang (2007) Optimizing Routing Metrics for Large-Scale Multi-Radio Mesh Networks. In WiCOM

[52] SartajSahni and Teofilo Gonzalez (1976) NP-complete approximation problems, Journal of the Association for Computing Machinery, Vol. 23.

[53] Das, A.K.; Vijayakumar, R.; Roy, S (2006) WLC30-4: Static Channel Assignment in Multi-radio Multi-Channel 802.11 Wireless Mesh Networks: Issues, Metrics and Algorithms,In IEEE Globecom.

[54] PradeepKyasanur, ChandrakanthChereddi, Nitin H. Vaidya (2006) Net-X: System eXtensions for Supporting Multiple Channels, Multiple Interfaces, and Other Interface Capabilities. UIUC Technical Report.

[55] Xiaoyan Hong, BoGu, Mohammad Hoque, Lei Tang (2010) Exploring multiple radios and multiple channels in wireless mesh networks, IEEE Wireless Communications, Vol. 17.

[56] Mohammad Hoque, Xiaoyan Hong, FarhanaAfroz (2009)Multiple Radio Channel Assignment Utilizing Partially Overlapped Channels, In IEEEGlobecom.

[57] Mohammad A Hoque, Xiaoyan Hong, (2009)Interference Minimizing Channel Assignment Using Partially Overlapped Channels in Multi-radio Multi-channel Wireless Mesh Networks, IEEE Southeast Conference.

[58] P.Guptaand P. R. Kumar(2000)The Capacity of Wireless Networks, IEEE Trans. on Information Theory, Vol. 46, No. 2.

[59] Mohammad A. Hoque, Xiaoyan Hong, Md. Ashfakul Islam, KaziZunnurhain (2010) "Delay analysis of Wireless Ad Hoc networks: Single vs. multiple radio," In Proceedings of IEEE LCN.

[60] K. Shih, C. Chang, D. Deng, and H. Chen (2010) Improving Channel Utilization by Exploiting Partially OverlappingChannels in Wireless Ad Hoc Networks,In IEEE Globecom.

[61] Yuting Liu, R. Venkatesan, and Cheng Li (2010) Load-Aware Channel Assignment Exploiting Partially Overlapping Channels for Wireless Mesh Networks,In IEEE Globecom.

[62] Yuting Liu, R. Venkatesan, and Cheng Li (2009) Channel assignment exploiting partially overlapping channels for wireless mesh networks,In IEEE Globecom.

[63] Pedro B. F. Duarte, Zubair Md. Fadlullah, Kazuo Hashimoto, and Nei Kato (2010) Partially Overlapped Channel Assignment on Wireless Mesh Network Backbone, In IEEE Globecom.

[64] Abbasi, S.,Kalhoro, Q., Kalhoro, M.A (2011) Efficient Use of Partially Overlapped Channels in 2.4 GHz WLAN Backhaul Links, International Conference on Innovations in Information Technology.

[65] http://www.cisco.com/en/US/docs/solutions/Enterprise/Mobility/emob30dg/RFDesign.html.

4

Analysis of WiMAX Physical Layer Using Spatial Multiplexing

Pavani Sanghoi[#1], Lavish Kansal[*2],

[#1]Student, Department of Electronics and Communication Engineering,
Lovely Professional University, Punjab, India
[1] pavani.sanghoi@gmail.com
[*2]Assistant Professor, Department of Electronics and Communication Engineering,
Lovely Professional University, Punjab, India
[2] lavish.s690@gmail.com

Abstract: *Broadband Wireless Access (BWA) has emerged as a promising solution for providing last mile internet access technology to provide high speed internet access to the users in the residential as well as in the small and medium sized enterprise sectors. IEEE 802.16e is one of the most promising and attractive candidate among the emerging technologies for broadband wireless access. The emergence of WiMAX protocol has attracted various interests from almost all the fields of wireless communications. MIMO systems which are created according to the IEEE 802.16-2005 standard (WiMAX) under different fading channels can be implemented to get the benefits of both the MIMO and WiMAX technologies. In this paper analysis of higher level of modulations (i.e. M-PSK and M-QAM for different values of M) with different code rates and on WiMAX-MIMO system is presented for Rayleigh channel by focusing on spatial multiplexing MIMO technique. Signal-to Noise Ratio (SNR) vs Bit Error Rate (BER) analysis has been done.*

Keywords: *BWA, WiMAX, OFDM, ISI, LOS, NLOS, MIMO, SNR, BER, FEC, CC*

I. INTRODUCTION

Worldwide Interoperability for Microwave Access (WiMAX) is an IEEE 802.16 standard based technology responsible for bringing the Broadband Wireless Access (BWA) to the world as an alternative to wired broadband. The IEEE 802.16e air interface standard [1] is basically based on technology namely, orthogonal frequency-division multiplexing (OFDM), that has been regarded as an efficient way to combat the inter-symbol interference (ISI) for its performance over frequency selective channels for the broadband wireless networks. The WiMAX standard 802.1 6e provides fixed, nomadic, portable and mobile wireless broadband connectivity without the need for direct line-of-sight with the base station. WiMAX distinguishes itself from the previous versions of the standard in the sense that this standard adds mobility to the wireless broadband standard.

WiMAX can be classified into Fixed WiMAX [2] and Mobile WiMAX. Fixed WiMAX is based upon Line Of Sight (LOS) condition in the frequency range of 10-66GHz whereas Mobile WiMAX is based upon Non-Line of Sight (NLOS) condition that works in 2-11 GHz frequency range [3]. For 802.16e standard, MAC layer & PHY layer has been defined, but in this paper, emphasis is given only on the PHY layer. PHY layer for mobile WiMAX which is IEEE-802.16e standard [4] has scalable FFT size i.e. 128-2048 point FFT with OFDMA, Range varies from 1.6 to 5 Km at 5Mbps in 5MHz channel BW, supporting 100Km/hr speed.

Multi-Input Multi-Output (MIMO) technology has also been renowned as an important technique for achieving an increase in the overall capacity of wireless communication systems. In this multiple antennas are employed at the transmitter side as well as the receiver side [5].

One can achieve spatial multiplexing gain in MIMO systems realized by transmitting independent information from the individual antennas, and interference reduction. The enormous values of the spatial multiplexing or capacity gain achieved by MIMO Spatial multiplexing technique had a major impact on the introduction of MIMO technology in wireless communication systems.

The paper is organized as follows: Model of WiMAX PHY layer is explained in section II. An overview of the MIMO systems is presented in Section III. MIMO Techniques are provided in section IV. WiMAX-MIMO systems are studied in section V. Results and simulations are shown in section VI. At the end conclusion is given in section VII.

II. WIMAX MODEL FOR PHYSICAL LAYER

The main task of the physical layer is to process data frames delivered from upper layers to a suitable format for the wireless channel. This task is done by the following processing: channel estimation, FEC (Forward Error Correction) coding, modulation, mapping in OFDMA (Orthogonal Frequency Division Multiple Access) symbols, etc [6]. The block diagram for WiMAX system (802.16e standard) is shown in Figure 1.

A. Randomization

It is the first process which is carried out in the WiMAX Physical layer after the data packet is received from the higher layers and each of the burst in Downlink as well as in the Uplink is randomized. It is basically scrambling of data to generate random sequence in order to improve coding performance and data integrity of the input bits [7].

B. Forward Error Correction (FEC)

It basically deals with the detection and correction of errors due to path loss and fading that leads to distortion in the signal. There are number of coding systems that are involved in the FEC process like RS codes, convolution codes, Turbo codes, etc. Basically we will be focusing upon the RS as well as the convolution codes.

1. RS codes

These are non-binary cyclic codes that add redundancy to the data. This redundancy is basically addition of parity bits into the input bit stream that improves the block errors.

2. Convolution codes (CC)

These CC codes introduce redundant bits in the data stream with the use of linear shift registers (m). The information bits are applied as input to shift register and the output encoded bits are obtained with the use of modulo-2 addition of the input information bits. The contents of the shift register in 802.11a physical layer uses Convolution code as the mandatory FEC [8]. These convolutional codes are used to correct the random errors and are easy to implement than RS codes. Coding rate is defined as the ratio of the input bits to the output bits. Higher rates like 2/3 and 3/4, are derived from it by employing "puncturing." Puncturing is a procedure that involves omitting of some of the encoded bits in the transmitter thus reducing the number of transmitted bits and hence increasing the coding rate of the CC code and inserting a dummy "zero" metric into the convolution viterbi decoder on the receive side of WiMAX Physical layer in place of the omitted bits. Code rate of convolution encoder is given as:-

$$\text{Code rate} = k/n$$

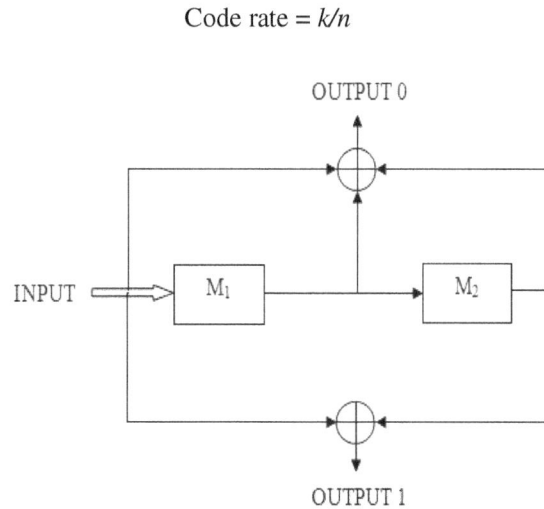

Fig. 1. Convolution Encoder; code rate=1/2, m=2

For Decoding the Viterbi algorithm is used at the receiver side of the PHY layer. To describe a convolution code, one need to characterize the encoding function (m), so that given an input sequence m, one can readily compute the output sequence U.

C. Interleaving

It aims at distributing transmitted bits in time or frequency domain or both to achieve desirable bit error distribution after the demodulation process. In interleaving, data is mapped onto non-adjacent subcarriers to overcome the effects of multipath distortion and burst errors. Block interleaving mainly operates on one of the block of bits at a time. The number of bits in each block is known as interleaving depth, which defines the delay introduced by interleaving process at the transmitter side. A block interleaver can be described as a matrix to which data is written in column format and data is read in row wise format, or vice versa.

D. Modulation

This process involves mapping of digital information onto analog form such that it can be transmitted over the channel. A modulator is involved in every digital communication system that performs the task of modulation. Modulation can be done by changing the amplitude, phase, as well as the frequency of a sinusoidal carrier. In this paper we are concerned with the digital modulation techniques. Various digital modulation techniques can be used for data transmission, such as M-PSK and M-QAM, where M is the number of constellation points in the constellation diagram. Inverse process of modulation called demodulation is done at the receiver side to recover the original transmitted digital information.

E. Pilot Insertion

Used for channel estimation & synchronization purpose. In this step, pilot carriers are inserted whose magnitude and phase is known to the receiver.

F. Inverse Fast Fourier Transform (IFFT)

An Inverse Fast Fourier transform converts the input data stream from frequency domain to time domain representing OFDM Subcarrier as the channel is basically in time domain. IFFT is useful for OFDM system as it generates samples of a waveform with frequency components satisfying the orthogonality condition such that no interference occurs in the subcarriers.

Similarly FFT converts the time domain to frequency domain as basically we have to work in frequency domain [9]. By calculating the outputs simultaneously and taking advantage of the cyclic properties of the multipliers FFT techniques reduce the number of computations to the order of $N \log N$. The FFT is most efficient when N is a power of two.

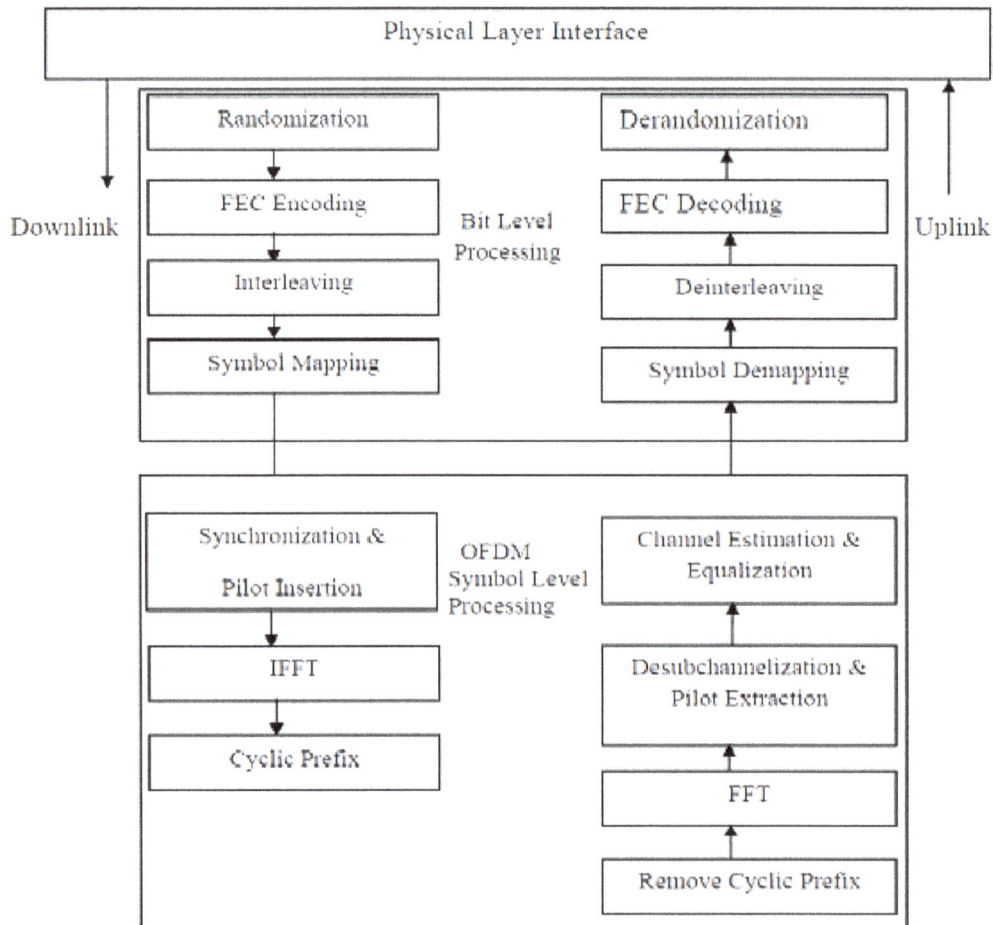

Fig. 2. WiMAX Model for Physical Layer (802.16)

G. Cyclic Prefix

One way to prevent ISI is basically to create a cyclically extended guard interval in between the data bits, where each of the OFDM symbol is preceded by a periodic extension of the signal itself which is known as the Cyclic Prefix as shown in fig. 3. When the guard interval is longer than the channel impulse response, or the multipath delay, the IS1 can be eliminated.

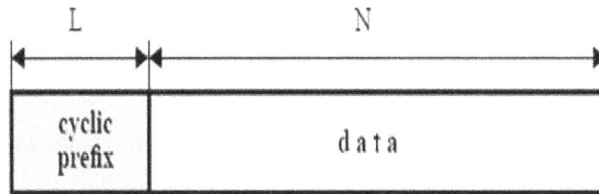

Fig. 3. Cyclic Prefix

H. Communication Channels

Communication channels are kind of medium of communication between transmitter and receiver. These channels are mainly divided into fast and slow fading channels. A channel is known as fast fading if the impulse response of the channel changes approximately at the symbol rate of the communication system, whereas in slow fading channel, impulse response stays unchanged for several symbols.

Rayleigh channel is used for the analysis purpose in this paper. Constructive and destructive nature of multipath components in flat fading channels can be approximated by Rayleigh distribution if there is no line of sight which means when there is no direct path between transmitter and receiver [10]. The received signal can be simplified to:

$$r(t) = s(t)*h(t) + n(t)$$

where h (t) is the random channel matrix having Rayleigh distribution and n(t) is the Additive White Gaussian noise. The Rayleigh distribution is regarded as the magnitude of the sum of two equal independent orthogonal Gaussian random variables and is useful in LOS condition only.

I. Receiver Side

The inverse processes take place at the receiver side. Removing the guard interval becomes equivalent to removing the cyclic prefix. Performing a FFT on the received samples after the cyclic prefix is discarded; the periodic convolution is transformed into multiplication, as it was the case for the analog Multi Carrier receiver [11]. Then demodulation, deinterleaving as well as FEC decoding using Viterbi Decoder and at last derandomization.

III. Multi Input Multi Output (MIMO) Systems

MIMO is an antenna technology for wireless communications in which multiple antennas are used at both the source (transmitter) and the destination (receiver). The antennas at each end of the communications circuit are combined to minimize errors and optimize data speed. Multi-antenna systems can be classified into three main categories. Multiple antennas used at the transmitter side of MIMO systems are mainly used for beamforming purposes to avoid the signal going to undesired directions. Transmitter or the receiver side multiple antennas are used for realizing different (frequency, space) diversity schemes in order to get diversity or capacity gain. The third class includes systems with multiple transmitter and receiver antennas that help in realizing spatial multiplexing which is often referred as MIMO by itself.

In radio communications, MIMO means employing multiple antennas at both the transmitter and receiver side of a specific radio link [12]. In case of spatial multiplexing technique, different data symbols are transmitted through different antennas with the same frequency within the same time interval. Multipath propagation phenomenon is assumed in order to ensure that there is correct operation of spatial multiplexing, since MIMO performs better in terms of channel capacity in a rich multipath scattering environment.

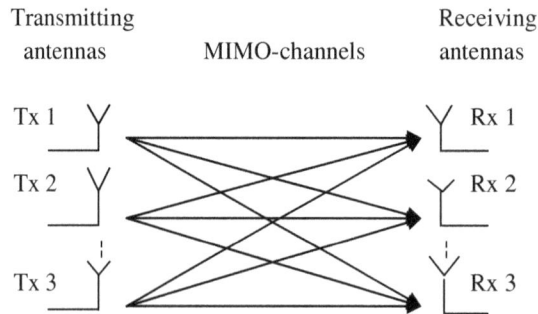

Fig. 4. Block Diagram of MIMO system with M Transmitters and N Receivers

IV. MIMO Techniques

A. Spatial Diversity

Diversity is one of the ways to combat the phenomenon of multipath fading. The main idea behind the diversity technique is to provide multiple replicas of the transmitted signal to the receiver. If these multiple replicas fade independently with each other, there is less probability for having all copies of the transmitted signal in deep fade simultaneously [13]. MIMO system takes advantage of the spatial diversity technique by placing independent separate antennas in a dense multipath scattering environment. In spatial diversity, same information is being sent on the independent individual antennas at the transmitter side. These systems can be implemented in a number of ways to obtain a diversity gain to combat with signal fading or a capacity improvement can also be done. Thus, the receiver can easily decode the transmitted signal using these received signals. This technique involves Space Time Block Coding (STBC) and Space Time Trellis Coding (STTC). Diversity techniques can be implemented into different ways in order to improve the bit error rate of the system [14].

B. Spatial Multiplexing

This form of MIMO is used to provide additional data capacity by utilizing the different paths to carry additional traffic, i.e. increasing the data throughput capability. The spatial multiplexing mitigates the multipath propagation phenomenon problem that is experienced by most of the microwave transmissions. MIMO techniques permit multiple streams that help in improving the signal-to-noise ratio and also the reliability that significantly improves over other versions of the standard. Spatial multiplexing includes transmitting different information onto different independent individual antennas at the transmitter side of the MIMO system and thus helps in attaining the capacity gain. This technique includes V-BLAST technology that is used to improve the spectral efficiency of the system [15].

MIMO systems utilize spatial multiplexing under rich scattering environment; independent data streams are simultaneously transmitted over different antennas to increase the effective data rate. MIMO spatial multiplexing [16] requires at least 2 transmitters and 2 receivers, and the receivers must be in the same place that means they should be in the same device. Because the transmitting antennas are not required to be in the same device and also two mobiles can be used together in the uplink.

C. Beamforming

Beamforming enables performance gains with multiple antennas at the BS and even a single antenna at the MS so as to direct the beam in a particular direction such that the signal going in the desired direction is increased and the signal going to the other directions is decreased. Such performance gains are derived from the array gain obtained from the phased array antennas used in the beamforming plus the diversity gain, which can be as much as 10 dB for a system having four antennas at the base station and a single antenna at the mobile station. Beamforming is a signal processing technique used to control the directionality of the transmission and reception of radio signals. The most effective and efficient type of beamforming is dynamic digital beamforming [17]. This uses an advanced, on-chip digital signal processing (DSP) algorithm in order to gain complete control over all the Wi-Fi signals.

V. WiMAX-MIMO Systems

MIMO systems created according to the IEEE 802.16-2005 standard (WiMAX) under different fading channels can be implemented to get the benefits of both the MIMO and WiMAX technologies. Main aim of combining both WiMAX and Spatial multiplexing MIMO technique is to achieve higher data rates by lowering the BER and improving the SNR of the whole system. The proposed block diagram of WiMAX-MIMO systems is given in Figure 5.

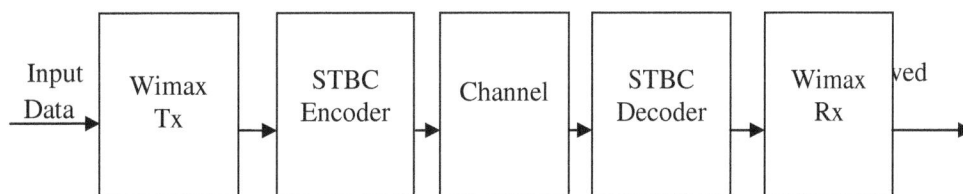

Fig. 5. WiMAX-MIMO System

The use of WiMAX technology with the MIMO technology provides an attractive solution for future broadband wireless systems that require reliable, efficient and high-rate data transmission. Employing MIMO systems in WiMAX [18] yields better BER performance compared to simple WiMAX protocol. Spatial multiplexing technique of MIMO systems provides spatial multiplexing gain that has a major impact on the introduction of MIMO technology in wireless systems thus improving the capacity of the system. Combining of both the systems involves employing STBC encoder and decoder at the transmitter and receiver side of WiMAX Physical Layer respectively.

This paper analyze the WiMAX protocol as well as the spatial multiplexing technique of MIMO systems in order to achieve higher data rates by lowering the Bit Error Rate and improving the SNR value of the system to achieve better performance and results. Spatial multiplexing (SM) is a recently developed transmission technique that uses multiple antennas and helps in achieving the capacity gain.

VI. Results and Simulations

In this paper behavior of the WiMAX-MIMO system under different modulations with different convolutional code rates is studied and the effects of increasing the order of the modulation on the BER performance of the system are presented. Results are shown in the form of SNR vs BER plot for different modulations that shows an improvement in the SNR value when we employ spatial multiplexing technique of MIMO system with the WiMAX protocol. In the case of spatial multiplexing, capacity improvement can also be seen as we are sending different data over the independent individual antennas. The analysis has been done for Rayleigh channel.

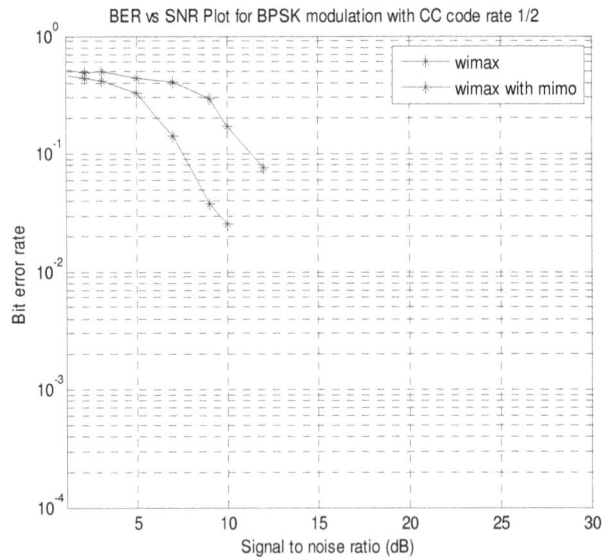

(a) BPSK with CC code rate 1/2

In this graph we are able to get 2dB improvement in SNR and capacity improvement is also shown when we employ SM technique of MIMO in WiMAX in the presence of Rayleigh channel.

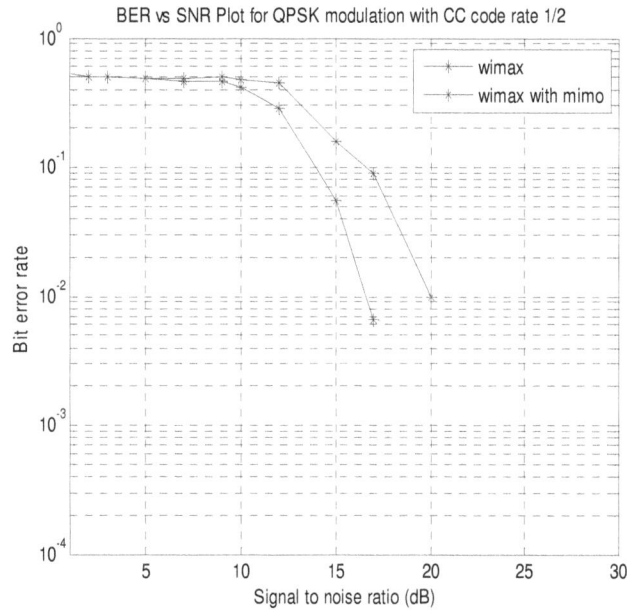

(b) QPSK with CC code rate 1/2

QPSK modulation with CC code rate 1/2 provides 2dB improvement in the SNR value when we combine WiMAX with MIMO SM technique as well as capacity gain can also be seen. Rayleigh channel is used for the analysis.

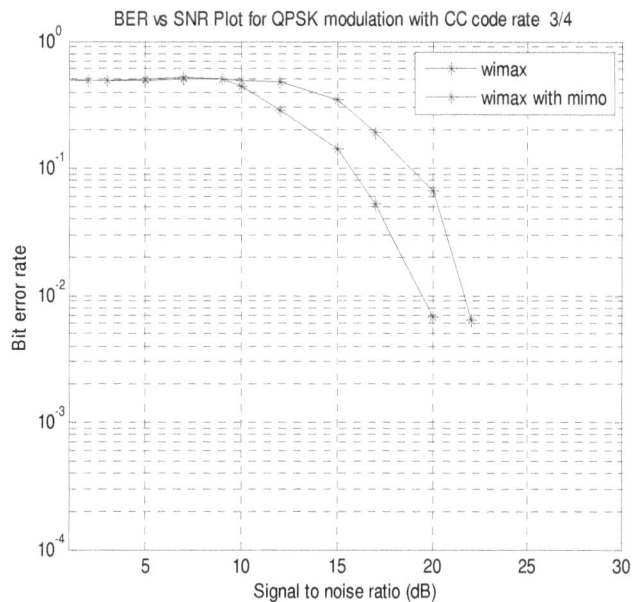

(c) QPSK with CC code rate 3/4

Capacity as well as SNR improvement of 2dB can be seen when we use QPSK modulation with CC code rate of 3/4 in WiMAX-MIMO system compared to simple WiMAX using Rayleigh channel.

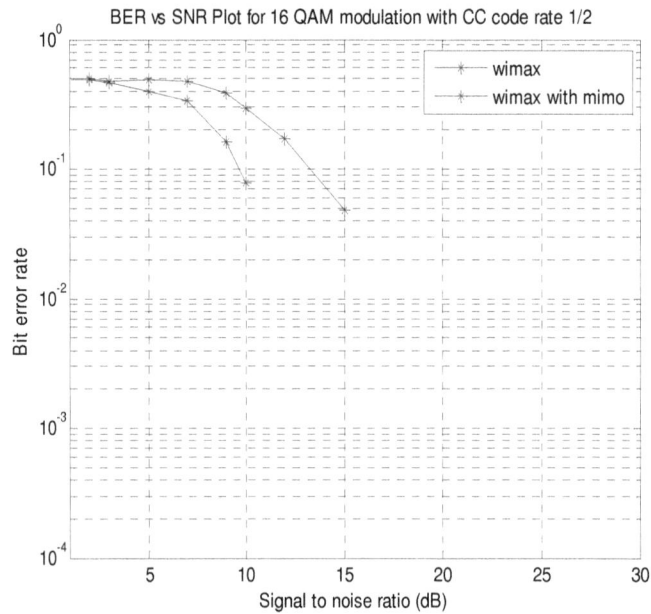

(d) 16QAM with CC code rate 1/2

As seen from the graph, SNR improvement of 5dB and capacity improvement is there using WiMAX with SM technique of MIMO technology in the presence of Rayleigh channel.

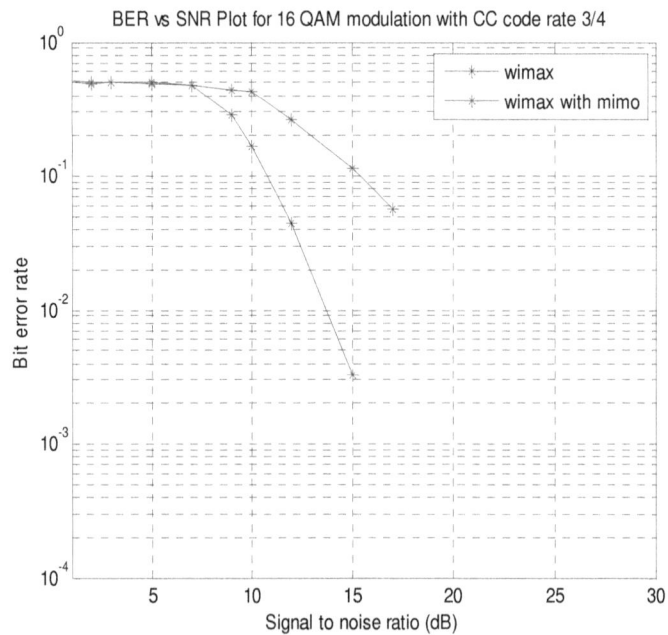

(e) 16 QAM with CC code rate 3/4

Again as we can see there is an improvement of 2dB in SNR using Rayleigh channel with combined effect of WiMAX-MIMO technology.

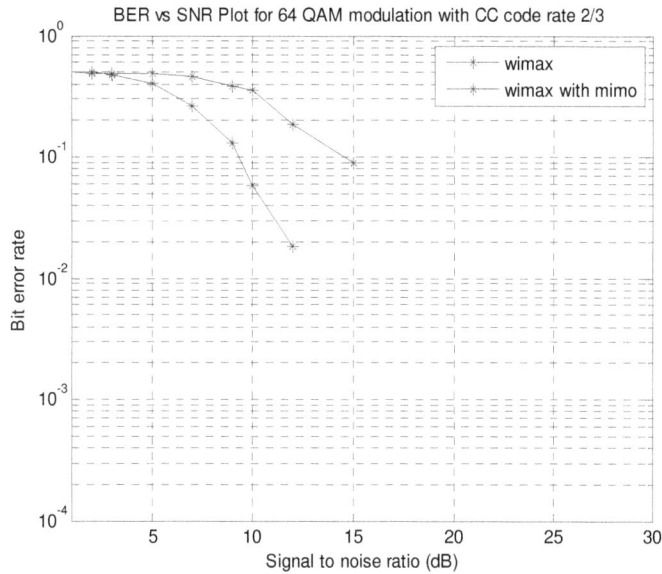

(f) 64 QAM with CC code rate 2/3

SNR improvement of 3dB in the case of 64QAM with CC code rate 2/3 can be seen and capacity improvement can also be seen in when we use WiMAX-MIMO in the presence of Rayleigh channel.

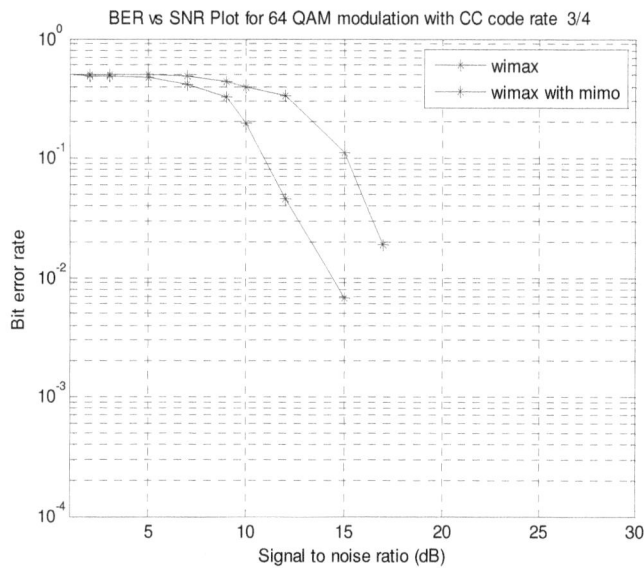

(g) 64 QAM with CC code rate 3/4

This graph shows an SNR improvement of 2dB as well as capacity gain using WiMAX-MIMO system in the presence of Rayleigh channel.

Fig. 6. BER vs SNR plots for Rayleigh channel
a) BPSK code rate 1/2 b) QPSK code rate 1/2 c) QPSK code rate 3/4 d) 16 QAM code rate 1/2

e) 16 QAM code rate 3/4 (f) 64 QAM code rate 2/3 (g) 64 QAM code rate 3/4

The performance in the form of BER vs SNR plots for different modulations over Rayleigh channel for WiMAX-MIMO system with different CC code rates have been presented in Figure 6 (a)–(g). Each graph shows an improvement in SNR using spatial multiplexing technique of MIMO system which is given in the following table.

Table 1: SNR improvement in Rayleigh channel by using Spatial Multiplexing in WiMAX

MODULATION	SNR Improvement using Rayleigh channel (dB)
BPSK code rate 1/2	2dB
QPSK code rate 1/2	3dB
QPSK code rate 3/4	2dB
16 QAM code rate 1/2	5dB
16 QAM code rate 3/4	2dB
64QAM code rate 2/3	3dB
64QAM code rate 3/4	2dB

VII. CONCLUSIONS

In this paper effect of employing spatial multiplexing technique of MIMO system in WiMAX 802.16e PHY layer has been simulated through Matlab2009a. This technique of MIMO systems provides spatial multiplexing gain that has a major impact on the introduction of MIMO technology in wireless systems. Rayleigh channel has been taken into account for the analysis purpose. Simulations are based upon using different modulations with different convolutional code rates and show that there is improvement in the SNR value as well as capacity improvement can also be seen by employing spatial multiplexing technique of MIMO system in WiMAX protocol.

Results are presented in the form of BER vs SNR value and show that BER reduces when we employ MIMO system in WiMAX in comparison to simple WiMAX. This shows that employing MIMO system in WiMAX improves the overall performance of the system and provides capacity gain. Main aim is to reduce the BER of the system for lower value of SNR hence providing higher data rates for the transmission purpose such that originality of the input signal is retained.

VIII. REFERENCES

[1] IEEE standard for Local and Metropolitan Area Networks-Part 16: Air interface for Fixed Broadband Wireless Access Systems, IEEE Std. 802.16, Oct. 2004

[2] IEEE standard 802.16-2005, IEEE standard for Local and Metropolitan Area Networks-Part16: Air Interface for Fixed and Mobile Broadband wireless Access system, Feb 2006.

[3] IEEE 802.16 WG,"IEEE Standard for Local and Metropolitan Area Network Part 16: Air Interface for Fixed Broadband Wireless Access Systems" IEEE Std 802.16-2004 p.1 - p.857

[4] IEEE 802.16WG,"IEEE standard for local and metropolitan area networks part 16: Air interface for fixed and mobile broadband wireless access systems, Amendment 2," IEEE 802.16 Standard, December 2005.

[5] Muquet, E. Biglieri, A. Goldsmith and H. Sari, "MIMO Techniques for Mobile WiMAX Systems", SEQUANS Communications White Paper, pp 1-18, 2006.

[6] W. A. C. Fernando, R.M.A.P. Rajatheva and K. M. Ahmed, "Performance of Coded OFDM with Higher Modulation Schemes", International Conference on Communication Technology, Vol. 2, pp 1-5, 1998.

[7] M. N. Khan, S. Ghauri, "The WiMAX 802.16e Physical Layer Model", International Conference on Wireless, Mobile and Multimedia Networks, pp 117-120, 2008.

[8] S. Bansal, R. Upadhyay, "Performance Improvement of Wi-Max IEEE 802.16e in Presence of Different FEC Codes", First International Conference on Computational Intelligence, Communication Systems and Networks, IEEE Computer Society, pp 226-229, 2009.

[9] M. Wang, "WiMAX Physical Layer: Specifications Overview and Performance Evaluation", 2nd IEEE CCNC Research Student Workshop, pp 10-12, 2011.

[10] J. Mountassir, H. Balta, M. Oltean, M. Kovaci & A. Isar, "A Physical Layer Simulator for WiMAX in Rayleigh Fading Channel", 6th IEEE International Symposium on Applied Computational Intelligence and Informatics , pp 281-281, 2011

[11] M. Oltean, M. Kovaci, J. Mountassir, A. Isar and P. Lazar, A physical layer simulator for WiMAX, *Proc. of the 9th IEEE International Symposium of Electronics and Telecommunications,* ISETC 2010, Timisoara, Romania, Nov. 2010, pp. 133-136.

[12] S. Alamouti, "A simple transmit diversity technique for wireless communications", IEEE Journal on Selected Areas of Communication, Vol. 16, pp 1451–1458, Oct. 1998.

[13] H. Hourani, "An overview of diversity techniques in wireless communication systems," IEEE, S-72.333 Postgraduate Course in Radio Communications, pp. 1-5, October 2004/2005.

[14] V. Tarokh, H. Jafarkhani and A. R. Calderbank, "Space–time block codes from orthogonal designs", IEEE Transactions on Information Theory, Vol. 45, pp 1456–1467, July 1999.

[15] P. W. Wolniansky, G. J. Foschini, G. D. Golden and R. A. Valenzuela, "V-Blast: An architecture for realizing very high data rates over the rich-scattering channel", International Symposium on Signals, Systems and Electronics, pp 295–300, 1998.

[16] R. Y. Mesleh, H. Haas, S. Sinanovic, C. W. Ahn and S. Yun, "Spatial modulation", IEEE Transaction on Vehicular Technology, Vol. 57, Issue 4, pp 2228-2241, July 2008.

[17] J. Huang, J. Zhang, Z. Liu, J. Li and X. Li, "Transmit Beamforming for MIMO-OFDM Systems with Limited Feedback", IEEE Vehicular Technology Conference, pp 1-5, 2008.

[18] O. Arafat, K. Dimyati, "Performance Parameter of Mobile WiMAX: *A Study on the Physical Layer of Mobile WiMAX under Different Communication Channels & Modulation Technique",* Second International Conference on Computer Engineering and Applications, pp 533-537, 2010.

5

STOCHASTIC ANALYSIS OF RANDOM AD HOC NETWORKS WITH MAXIMUM ENTROPY DEPLOYMENTS

Thomas Bourgeois[1] and Shigeru Shimamoto[1]

[1]Graduate School of Global Information and Telecommunication Studies
Waseda University, Japan.

ABSTRACT

In this paper, we present the first stochastic analysis of the link performance of an ad hoc network modelled by a single homogeneous Poisson point process (HPPP). According to the maximum entropy principle, the single HPPP model is mathematically the best model for random deployments with a given node density. However, previous works in the literature only consider a modified model which shows a discrepancy in the interference distribution with the more suitable single HPPP model. The main contributions of this paper are as follows. 1) It presents a new mathematical framework leading to closed form expressions of the probability of success of both one-way transmissions and handshakes for a deployment modelled by a single HPPP. Our approach, based on stochastic geometry, can be extended to complex protocols. 2) From the obtained results, all confirmed by comparison to simulated data, optimal PHY and MAC layer parameters are determined and the relations between them is described in details. 3) The influence of the routing protocol on handshake performance is taken into account in a realistic manner, leading to the confirmation of the intuitive result that the effect of imperfect feedback on the probability of success of a handshake is only negligible for transmissions to the first neighbour node.

KEYWORDS

*Ad Hoc Networks · Point Process · Stochastic Geometry ·
Medium Access Control*

1. INTRODUCTION

An ad hoc wireless network consists of self-organizing transceivers communicating with one another in a decentralized way. Such network is referred to as random when the location of nodes in the d-dimensional Euclidean space $\mathbb{R}^d$ is statistically random [6,21,1]. Using tools from stochastic geometry, the mathematical analysis of carefully chosen models can shed light on the behaviour of a Randomly distributed Ad Hoc Network (RAHN) and can provide insights into the design of Medium Access Control (MAC) and routing protocols [22,23]. The most popular spatial distribution used up to now to model large RAHNs has been the homogeneous Poisson Point Process (HPPP). In a HPPP of intensity λ, the number of nodes $N(B)$ in a given area $B \subset \mathbb{R}^d$ is Poisson distributed with mean $\lambda |B|$, where |.| denotes the d-dimensional volume. More point processes and their applications are described in [2], however in this work we focus our analysis on HPPP-based deployments. The popularity of the HPPP is mainly due to its tractability. A less often mentioned fact is that it has maximum entropy among all point processes of a given average rate parameter (i.e, density λ) [8]. Indeed, from the principle of maximum entropy, and with no

prior knowledge about the distribution of nodes in the network, it is best to choose a model with a minimum of prior information built in [15,16]. Consequently,
in an ad hoc network using the "single HPPP" model, hence the motivation for the present work.

The main contributions of this paper are as follows. 1) It presents a new mathematical framework leading to closed form expressions of the probability of success of both one-way transmissions and handshakes for a deployment modelled by a single HPPP. Our approach, based on stochastic geometry, can be extended to complex protocols. 2) From the obtained results, all confirmed by comparison to simulated data, optimal PHY and MAC layer parameters are determined and the relations between them is described in details. 3) The influence of the routing protocol on handshake performance is taken into account in a realistic manner, leading to the confirmation of the intuitive result that the effect of imperfect feedback on the probability of success of a handshake is only negligible for transmissions to the first neighbor node.

Parts of this work were presented in [7]. However, the present paper provides complete, revisited proofs, more detailed explanations as well as additional simulation results. Throughout the paper, we aim to obtain the probability of success of a transmission (one-way and handshake) to the k-th nearest neighbour in an ad hoc network modelled by a single HPPP in which nodes employ Slotted-ALOHA (without and with Acknowledgement, respectively). The remainder of this paper is organized as follows. In Section 2, we present our general system model and some preliminary results regarding interference modelling in a network modelled by a single HPPP. In section 3, exploiting results from section 2, we calculate the transmission success probability between two arbitrarily chosen nodes in the network. Reasonable approximations are used to render the analysis tractable. In section 4, we apply the results of the sections 2 and 3 to the cases of the MAC protocols S-ALOHA and S-ALOHA with ACK packet. In section 5, we validate the approximate theoretical expressions obtained in the previous section by comparing them to computer simulation results and analyse the conditions for their validity. Finally, in section 6, a conclusion summarizes the paper.

2. SYSTEM MODEL AND PRELIMINARY RESULTS

2.1. Deployment and channel models

In the remainder, independently of the protocols considered and before specific roles (i.e, transmitter or receiver) are attributed, node deployment is always assumed to be modelled by a d-dimensional HPPP with intensity λ. The channel model incorporates path-loss and fast fading. The path-loss will follow the unbounded power law $l\left(r\right) = r^{-\alpha}$ where α is called the path-loss coefficient. The fading is assumed to be Rayleigh distributed, the square thereof then following an exponential distribution with rate parameter equal to one. Transmissions in the network will be slotted and the fading is assumed to be constant over the duration of a slot and independent from one slot to another. Moreover, we always assume fading coefficients of all links to be independent from one another, independent of node position and identically distributed. Since our goal is to analyse the influence of a large number of simultaneous transmissions on the network performances, we will also assume the network to be interference-limited. Hence, as in [12,4], a transmission will be considered successful if the received signal power S from the intended transmission is greater than the interference I generated by all other transmissions by a factor larger than a given threshold θ. That is, the inequality $\frac{S}{I} > \theta$ must be true for a transmission to be successful.

2.2. Interference model

Let us consider a given network node denoted as Rx. For any realization of the HPPP, we can sort other nodes by order of distance to Rx. Let us define for a given slot the random squared fading coefficient $G_i, i \in \{1, ..., +\infty\}$ of the channel between the i-th closest node and Rx, R_i the random distance from the i-th closest node to Rx. Furthermore, for all realizations of the HPPP, let us assume that the n-th closest node has a packet for Rx and that all nodes except Rx transmit during the slot. For a receiver threshold θ, the probability for this intended transmission to be successful is given by

$$
p_{rx}^n(\theta) = \mathbb{E}\left\{\mathbb{P}\left(G_n > \theta R_n^\alpha \sum_{i=1, i\neq n}^\infty G_i R_i^{-\alpha} \middle| \vec{R}_{1,\infty}, \vec{G}_{1,\infty\setminus n}\right)\right\} (1)
$$

where $p_{rx}^n(\theta)$ denotes the transmission success probability in a "receiver-centric" approach, with the n-th neighbour being the transmitter and $\mathbb{E}\{.\}$ is the expectation operator over the random vectors $\vec{R}_{1,\infty} = [R_1, ..., R_\infty]^T$ and $\vec{G}_{1,\infty\setminus n} = [G_1, ..., G_{n-1}, G_{n+1}, ..., G_\infty]^T$. Then, given that the Complementary Cumulative Distribution Function of an exponential variable with rate parameter one is $F^c(x) = e^{-x}$, we obtain from (1),

$$
p_S(\theta) = \mathbb{E}_{R_n, I}\left\{e^{-\theta R_n^\alpha I} | R_n, I\right\} = \mathbb{E}_{R_n}\{L_I(\theta R_n^\alpha) | R_n\} \qquad (2)
$$

where $I = \sum_{i=1, i\neq n}^\infty G_i R_i^\alpha$ is the interference term and $L_I(s) = \mathbb{E}_I\{e^{-sI}\}$ is its Laplace Transform. In the following subsections, we determine the Laplace Transform of the interference originating from two important node sets.

2.3. Interference from the n − 1 closest interferers

Keeping the scenario from the last subsection, consider the $n-1$ closest nodes to Rx. These nodes are uniformly distributed over a d-dimensional ball with random radius R_n deprived from its center. However, given that the volume occupied by the center point (where Rx is located) is null, we may as well consider the n − 1 nodes to be uniform over the whole ball. Thus, the n − 1 nodes form a BPP with random intensity $\lambda = \frac{n-1}{c_d R_n^d}$ where c_d indicates the volume of a d-dimensional ball with unit radius. In these conditions, we have the following property.

Proposition 1: Consider the annular region $\mathcal{A}(Rx, A, B)$ with center Rx, inner radius A and outer radius $B < R_n$. Given that there are exactly k interferers in the annular region $\mathcal{A}(Rx, A, B)$, the Laplace Transform of the interference they generate at Rx is given by:

$$
L_{I_{\mathcal{A}(Rx,A,B)}|k}(s) = \left(1 + \frac{B^d\Omega(s, B) - A^d\Omega(s, A)}{B^d - A^d}\right)^k (3)
$$

with

$$
\Omega(s, X) = \frac{F\left(1, 1, 1 - \frac{d}{\alpha}, \frac{sX^{-\alpha}}{1+sX^{-\alpha}}\right)}{1 + sX^{-\alpha}} - B\left(1 - \frac{d}{\alpha}, 1 + \frac{d}{\alpha}\right)s^{\frac{d}{\alpha}}X^{-d} - 1 (4)
$$

where $F(a, b, c, x)$ is the Gaussian hypergeometric function and $B(a, b)$ denotes the Beta function. To ease numerical computation, a simple approximation of $\Omega(s, X)$ is proposed in section IV.

Proof: See appendix A.

The Laplace Transform of the interference at Rx originating from all nodes in the ball, save the center, can be obtained by setting $B = R_n$, $k = n - 1$ and making A converge toward zero. Then, to obtain a closed form expression, one may use (3) and the following proposition.

Proposition 2: The limit of $A^d \Omega(s, A)$ for A converging toward zero is given by

$$\lim_{A \to 0^+} A^d \Omega(s, A) = 0 \quad (5)$$

Proof: See appendix B.

We note also that setting $s = \theta r^\alpha$, $A = ar$ and $B = br$ where $a, b > 0$ in (4) renders the Laplace Transforms in (3) independent of r, for any $r > 0$. This property will prove useful when considering (2). Hence, for notational convenience, we will define the following function. For any positive real numbers a and b such that $a \leq b$, $\Phi(\theta, a, b) = L_{I_{A(O, ar, br)|1}}(\theta r^\alpha)$ for all $r > 0$. Note that, for the sake of brevity, $\Phi(\theta, a, b)$ will be denoted as $\Phi(a, b)$ in the following when the threshold considered is clear from the context.

2.4. Interference from nodes beyond the n-th neighbour

Keeping the scenario presented in subsection 2.2, we now consider all nodes further away from Rx than the n-th closest node. In these conditions, we have the following property.

Proposition 3: The Laplace Transform of the interference generated by all nodes away from Rx by a distance greater than R_n is given by

$$L_{I_{\mathbb{R}^d / B(Rx, R_n)}}(s) = e^{-\lambda c_d R_n^d \left(\Omega(s, R_n) + B\left(1 - \frac{d}{\alpha}, 1 + \frac{d}{\alpha}\right) s^{\frac{d}{\alpha}} R_n^{-d} \right)} \quad (6)$$

Proof: See appendix C.

In the following sections, we present general expressions of transmission success probabilities for various scenarios in the single HPPP model.

3. Transmission Success Probability in an ad hoc network modelled as a single HPPP

3.1. Neighbour index definition

In this subsection, we introduce the notion of neighbour index. As mentioned briefly in introduction, although employing the PGFL leads to a simple expression of the transmission success probability, it is only valid for a given transmitter-receiver distance. However, in the case of a single HPPP, this distance is random with an unknown distribution (since both nodes are part of the original generating HPPP). Conditioning transmitter and receiver on their mutual neighbour indexes allows defining the distance distribution and thus enables averaged results to be inferred. The neighbour index is defined as follows.

Definition 1: For two distinct network nodes Tx and Rx, Tx has neighbour index $n \in \mathbb{N}^*$ with respect to Rx, which we denote $\mathcal{I}_{Rx}(Tx) = n$, if there exists exactly $n - 1$ nodes closer from Rx than Tx. If the identity of nodes is clear from the context, one may simply write $\mathcal{I} = n$.

From this definition, two different approaches can be considered.

In the receiver-centric approach, transmitters with a packet intended for a given receiver (namely, "intended transmitters", as opposed to "interferers") are conditioned on their neighbour index with respect to the considered receiver node. In the transmitter-centric approach, it is the receiver which is conditioned on its neighbour index with respect to the intended transmitter.

The following conditional probability mass function (p.m.f) $\mathcal{F}(n, k)$ will prove useful in order to relate both approaches. For Tx and Rx being two distinct network nodes, we define

$$\mathcal{F}(n, k) = \mathbb{P}\left(\mathcal{I}_{Tx}(Rx) = n \mid \mathcal{I}_{Rx}(Tx) = k\right) \quad (7)$$

When the identity of nodes is clear from the context, we will use simply $\mathcal{I}$ instead of $\mathcal{I}_{Tx}(Rx)$ and its symmetric $\mathcal{I}_{sym}$ instead of $\mathcal{I}_{Rx}(Tx)$ to denote neighbour indexes.

Proposition 4: For points taken from a d-dimensional HPPP , the p.m.f in (7)

is given by

$$\mathcal{F}(n, k) = \sum_{l=0}^{m-1} w_{n,k}(l) \quad (8)$$

where

$$w_{n,k}(l) = \binom{k-1}{l}\binom{n+k-l-2}{k-1} \frac{\beta_d^l (1 - \beta_d)^{n-k-2l-2}}{(2 - \beta_d)^{n+k-l-1}} \quad (9)$$

And $m = \min(n, k)$. Also, for $\|TxRx\| = x \in \mathbb{R}^{*+}$ and two d-dimensional balls $\mathcal{B}_d(Tx, x)$ and $\mathcal{B}_d(Rx, x)$, $\beta_d = \beta_d$ defined as $\beta_d = \frac{|\mathcal{B}(Tx,x) \cap \mathcal{B}(Rx,x)|}{|\mathcal{B}(Tx,x)|} = B_{\frac{3}{4}}\left(\frac{d+1}{2}, \frac{1}{2}\right)$ with $B_x(a, b)$ denoting the incomplete Beta function.

Proof: The proof for (8), if slightly incorrect, was given in [18]. Indeed,

in this reference, the author did not consider that the number of points in $\mathcal{B}(Tx, x) \cap \mathcal{B}(Rx, x)$ is at most $\min(n, k) - 1$. Nevertheless, this is a minor error which does not invalidate the thought process of the proof provided, hence the omission of its presentation in this paper. However, we provide in Appendix D the proof that the p.m.f as defined above does satisfy the normalization condition. *End of proof.*

Also, for notational convenience, we will define for the remainder $[y]_p = py + 1 - p$, which will prove useful when dealing with the expected value $\mathbb{E}\left\{y^X\right\}$ where $y \in \mathbb{R}$ and X is binomially distributed according to $X \sim \mathfrak{B}(N, p)$.

In the following subsections, we show how the preliminary results developed above may be used to obtain the transmission success probability in various contexts.

3.2. Receiver-centric approach for a single intended transmission

In this subsection, we consider the scenario described in subsection 2.2.

Theorem 1: Given that $\mathcal{I}_{Rx}(Tx) = n$, the probability of success for a transmission from Tx to Rx is given by

$$p_{rx}^n(\theta) = \frac{\Phi(0,1)^{n-1}}{\left(1 + \Omega(\theta x^\alpha, x) + B\left(1 - \frac{d}{\alpha}, 1 + \frac{d}{\alpha}\right)\theta^{\frac{d}{\alpha}}\right)^n} \quad (10)$$

Proof: The proof relies on the observation that, according to the independence of the number of nodes in disjoint areas in a HPPP, the transmission success probability can be written as follows.

$$p_{rx}^n(\theta) = \mathbb{E}_{R_n}\left\{\prod_{j=1}^{\infty} L_{I_{\mathcal{S}_j(R_n)}}(\theta R_n^\alpha)\right\} \quad (11)$$

where R_n is the distance from Rx to Tx and $\{\mathcal{S}_j(R_n)\}_{j=1,\ldots,\infty}$ is a complete partition of the d-dimensional space into disjoint areas functions of R_n and $I_{\mathcal{S}_j(R_n)}$ is the interference contribution from the nodes located in the area $\mathcal{S}_j(R_n)$. The final result is obtained by finding a suitable decomposition of the space.

The $n-1$ "close" interferers located in $\mathcal{B}(Rx, R_n)$ form an homogeneous BPP, while the "far" interferers (i.e, away from Rx by a distance greater than R_n) form an inhomogeneous PPP . Let us define the following regions: $\mathcal{S}_{n-} = \mathcal{B}(Rx, R_n), \mathcal{S}_{n+} = \overline{\mathcal{B}(Rx, R_n)}$. These two regions being disjoint, their node distributions are independent, thus the interference $I_{n-}(R_n)$ at Rx due to the "close" interferers and the interference $I_{n+}(R_n)$ at Rx due to the "far" interferers are independent. As a consequence, from (6) and from (3) for $A \to 0^+$, $B = R_n$ with $n-1$ nodes in $\mathcal{A}(Rx, A, B)$, we obtain that the success probability of a transmission from Tx to Rx can be expressed as

$$p_S(\theta | \mathcal{I}_{Rx}(Tx) = n) = \mathbb{E}_{R_n}\left\{L_{I_{n-}(R_n)}(\theta R_n^\alpha) L_{I_{n+}(R_n)}(\theta R_n^\alpha)\right\}$$
$$= \mathbb{E}_{R_n}\left\{\Phi(0,1)^{n-1} e^{-\lambda c_d R_n^d\left(\Omega(\theta R_n^\alpha, R_n) + B\left(1 - \frac{d}{\alpha}, 1 + \frac{d}{\alpha}\right)\theta^{\frac{d}{\alpha}}\right)}\right\} \quad (12)$$

From [18], the distance in a d-dimensional PPP from a node to its n-th nearest neighbour is governed by the probability density function (p.d.f)

$$f_{R_n}(r) = d\frac{(\lambda c_d)^n}{\Gamma(n)} r^{dn-1} e^{-\lambda c_d r^d} \quad (13)$$

Taking the expectation of (11) with respect to $R_n = \|TxRx\|$ thus leads to an integral of the form $\int_0^{+\infty} r^{dn-1} e^{-\mu r^d} dr$ where $\mu > 0$, which can be solved easily by applying the variable substitution $y = \mu r^d$ and then recognizing in the ensuing integral the Gamma function $\Gamma(n)$. The final result follows then easily, which completes the proof. *End of proof.*

3.3. Transmitter-centric approach for a single intended transmission

We now consider the transmitter-centric approach.

Theorem 2: The probability of a successful transmission from Tx to Rx, conditioned on $\mathcal{I}_{Tx}(Rx) = k$, can be approximated by[1]

$$p_{tx}^k(\theta) \approx \frac{\Phi(1,2)^{k-1}}{(1+K(\theta))^k} \mathbb{E}_{\mathcal{I}} \left\{ \Phi(0,1)^{\mathcal{I}-1} \left[\Phi(1,2)^{-1} \right]_{\beta_d}^{\min(\mathcal{I},k)-1} | \mathcal{I}_{sym} = k \right\} \qquad (14)$$

where $K(\theta) = 2^d \Omega(\theta x^\alpha, 2x) + B\left(1 - \frac{d}{\alpha}, 1 + \frac{d}{\alpha}\right)\theta^{\frac{d}{\alpha}} + \left(2^d - 2 + \beta_d\right)(1 - \Phi(1,2))$ where $p_{tx}^k(\theta)$ denotes the transmission in a "transmitter-centric" approach, with the k-th neighbour being the receiver.

Proof: Provided that $\mathcal{I}_{Tx}(Rx) = k$, $\mathcal{I}_{Rx}(Tx)$ is a random variable with distribution given by (7) and $R = \|TxRx\|$ is distributed according to (12) in which n is replaced by k. Under these conditions, we consider the following partition of $\mathbb{R}^d$.

$$\mathcal{S}_1(R) = \mathcal{B}_d(Rx, R) \cap \mathcal{B}_d(Tx, R)$$
$$\mathcal{S}_2(R) = \mathcal{B}_d(Rx, R) \cap \overline{\mathcal{B}_d(Tx, R)}$$
$$\mathcal{S}_3(R) = \overline{\mathcal{B}_d(Rx, R)} \cap \mathcal{B}_d(Tx, R)$$
$$\mathcal{S}_4(R) = \mathcal{A}(Rx, R, 2R) \cap \overline{\mathcal{S}_3}$$

The geometry of these zones is illustrated in figure 1. In the following, we simply denote $\mathcal{S}_j(R)$ as $\mathcal{S}_j$ for the sake of brevity.

We observe that $\mathcal{N}(\mathcal{S}_1)$ is a binomially distributed random variable $\mathfrak{B}(\mathcal{M}-1, \beta_d)$, where $\mathcal{M} = \min(\mathcal{I}_{Rx}(Tx), k)$. Partitioning the space using the above-defined regions, the transmission success probability can be expressed as

$$p_{tx}^k(\theta) = \mathbb{E}_{R, \mathcal{I}_{Rx}(Tx)} \left\{ \Phi(0,1)^{\mathcal{I}_{Rx}(Tx)-1} \mathbb{E}_{\mathcal{N}(\mathcal{S}_1)} \left\{ L_{I_{\mathcal{S}_3}}(\theta R^\alpha) \right\} \right.$$
$$\left. \mathbb{E}_{\mathcal{N}(\mathcal{S}_4)} \left\{ L_{I_{\mathcal{S}_4}}(\theta R^\alpha) \right\} e^{-\lambda c_d R^d \left(2^d \Omega(\theta R^\alpha, 2R) + B\left(1 - \frac{d}{\alpha}, 1 + \frac{d}{\alpha}\right)\theta^{\frac{d}{\alpha}}\right)} \right\} \qquad (15)$$

We note that $L_{I_{\mathcal{S}_3}}(\theta R^\alpha)$ is dependent on $\mathcal{N}(\mathcal{S}_1)$, through the relation $\mathcal{N}(\mathcal{S}_3) = k - 1 - \mathcal{N}(\mathcal{S}_1)$ while the remaining functions are only dependent on $\mathcal{I}_{Rx}(Tx)$ and R. From the properties of HPPPs, the number of nodes in $\mathcal{S}_4$ is Poisson distributed and independent of the number of nodes in other regions.

The distance from Rx to any node uniformly distributed in $\mathcal{S}_3$ (or in $\mathcal{S}_4$, respectively) follows a distribution with a non-trivial expression, rendering difficult any attempt to compute a closed form expression for the Laplace Transform of the interference contribution from nodes located in these regions. So as to keep the calculations tractable, the distribution of the distance from Rx to a node in the above-mentioned regions is approximated by the distribution of the distance from Rx to a node uniformly deployed over the annular region $\mathcal{A}(Rx, R, 2R)$.

A reasonable justification for this approximation is as follows. In order to obtain a simple approximation regarding a distribution, one may successively "forget" part of the constraints on the actual distribution and select the maximum-entropy distribution fitting the remaining

---

[1] The expectation in the expression of transmission success probability for the transmitter-centric approach is on the discrete random variable $\mathcal{I}$, which leads to an infinite sum. In practice, computing the first few terms produces values very close to the actual result.

constraints, if it exists. In our case, removing information regarding the shapes of $\mathcal{S}_3$ and $\mathcal{S}_4$, the only remaining knowledge we have is that both zones are inside $\mathcal{A}\left(Rx, R, 2R\right)$. However, the uniform distribution is the maximum entropy distribution among all continuous distributions supported in a bounded set, given that no other statistical characteristics are known about the actual distribution. Consequently, it is also the optimal distribution estimate for the actual distribution of nodes, according to the maximum entropy principle.

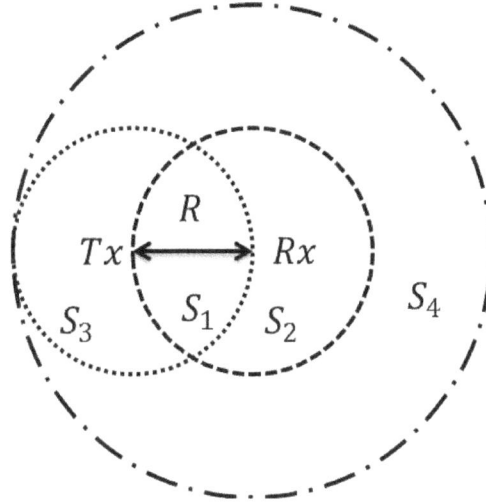

Figure 1 Illustration of the interference situation at a receiver Rx given an intended transmitter Tx located at distance R. The different zones in the figure are defined in (14).

As a result, the interference generated by a given number of nodes in $\mathcal{S}_3$ (or $\mathcal{S}_4$) may be approximated by the interference due to the same number of nodes uniformly distributed in $\mathcal{A}\left(Rx, R, 2R\right)$.

Using this approximation for $\mathcal{S}_3$, we obtain

$$\mathbb{E}_{\mathcal{N}(\mathcal{S}_1)}\left\{L_{I_{\mathcal{S}_3}}\left(\theta R^\alpha\right)\right\} \approx \sum_{i=0}^{\mathcal{M}-1} \binom{\mathcal{M}-1}{i} \beta_d^i (1-\beta_d)^{\mathcal{M}-1-i} \Phi\left(1,2\right)^{k-1-i}$$

$$= \Phi\left(1,2\right)^{k-1}\left(\frac{\beta_d}{\Phi\left(1,2\right)} + 1 - \beta_d\right)^{\mathcal{M}-1} \tag{16}$$

where the last line follows from the binomial theorem.

Regarding $\mathcal{S}_4$, given that $|\mathcal{S}_4| = \left(2^d - 2 + \beta_d\right)c_d R^d$, we have from the definition of the Poisson distribution that

$$\mathbb{E}_{\mathcal{N}(\mathcal{S}_4)}\left\{L_{I_{\mathcal{S}_4}}\left(\theta x^\alpha\right)\right\} \approx e^{-\lambda\left(2^d-2+\beta_d\right)c_d R^d} \sum_{j=0}^{+\infty} \frac{\left(\lambda\left(2^d-2+\beta_d\right)c_d R^d\right)^j}{j!} \Phi\left(1,2\right)^j$$

$$= e^{-\lambda c_d R^d\left(2^d-2-\beta_d\right)\left(1-\Phi(1,2)\right)} \tag{17}$$

Further taking the expectation over $\mathcal{I}_{R.r}(T.x)$ and then taking the expectation over R in (15) leads to an integral that can be solved in the exact same way as the one leading to (9), which completes the proof. *End of proof.*

In the next section, we show how results calculated up to now may be applied to the analysis of performances of some simple Medium Access Control protocols for ad hoc networks.

4. APPLICATION TO THE ANALYSIS OF SLOTTED-ALOHA

In this section, we analyse the success probability of transmissions in an ad hoc network employing the MAC protocols Slotted-ALOHA with and without ACK packet. The system model used in the following is as described in subsection 2.1.

4.1. Case of S-ALOHA with constant Medium Access Probability

In previous sections, we considered that all nodes in the network were active transmitters, except for a single receiver Rx. However, node activity in real networks is controlled by the MAC protocol.

In a Slotted-ALOHA (S-ALOHA) protocol with fixed Medium Access Probability (MAP) p, each node in the network with a packet to transmit does so during the next slot if it passes a random test with success probability p, and stays silent otherwise [14]. We assume in the following that all silent nodes can act as receiver. Let us further assume that nodes always have at least one packet ready for transmission and that consequently all nodes passing the random test do transmit. Then, let us consider an ad hoc network modelled as an HPPP $\mathcal{X}$ with constant intensity λ. In this case, employing an S-ALOHA protocol with constant MAP p results, from the thinning properties of PPPs [12], in the separation of the original HPPP into two distinct HPPPs. Namely, $\mathcal{X}_1$ with intensity $p\lambda$ and $\mathcal{X}_0$ with intensity $(1-p)\lambda$, which correspond to active transmitters and silent nodes, respectively.

The expressions of transmission success probability obtained in the previous section can be adapted easily to the case of Slotted-ALOHA with constant MAP by taking into account the following two observations. 1) The independent thinning of the original process $\mathcal{X}$ has for consequence that, for any given bounded set $\mathcal{S} \subset \mathbb{R}^d$, the number of active transmitters $\mathcal{N}_1(\mathcal{S})$ in $\mathcal{S}$ is binomially distributed (i.e $\mathcal{N}_1(\mathcal{S}) \sim \mathfrak{B}(\mathcal{N}(\mathcal{S}),p)$). 2) For unbounded regions of $\mathbb{R}^d$, active transmitters located in them form an inhomogeneous PPP with intensity $p\lambda$. It is easily verified that the Laplace Transform of their interference contribution may be obtained from the case treated in the previous section merely by changing λ into $p\lambda$ in the equations.

In the next subsection, we analyse the case of Slotted-ALOHA with Acknowledgement.

4.2. Case of constant MAP S-ALOHA with ACK slots

We now consider the same system as in the previous subsection, except that now each data transmission time slot is followed by a slot reserved for the transmission of ACK packets. Thus, nodes which receive successfully a packet intended for them during a data slot send an ACK packet during the following ACK slot.

Although interference in a HPPP is spatially correlated, we assume in the following that transmission successes on different links are independent so as to keep the derivations of handshake performances tractable. So as to show this assumption is reasonable, simulated data used for comparison will take into account the spatial correlation of interference.

An acknowledgement packet is considered successful if its SIR ratio is greater than the threshold θ'. If an intended data packet and the following ACK packet are both correctly received, then the handshake is considered successful. We assume that fading coefficients in data slots and ACK slots are independent.

Also, we define $q(\theta)$ the spatially averaged probability for any node in the network not transmitting during a given data slot to send a packet during the following ACK slot. That is, $q(\theta)$ can be seen as the spatially averaged probability that a node receives successfully a packet intended for itself during a given data slot, conditioned on its belonging to the receiver set. As a consequence, $q(\theta)$ not only depends on the MAP p, but also on the routing protocol employed (i.e, the way nodes in the network choose their packet destinations). For a given receiver $Rx \in \mathcal{X}_0$, the general expression for $q(\theta)$ is given by

$$q(\theta) = p \sum_{n=1}^{\infty} \mathbb{P}(X_n \to Rx)\, p_{rx}^n(\theta) \tag{1} \tag{18}$$

in which X_n denotes the n-th nearest node to Rx, $X_n \to Rx$ means " X_n selects Rx as destination" and where $p_{rx}^n(\theta)$ is the transmission success probability from X_n to Rx.

Consequently, provided that the routing protocol employed allows $\mathbb{P}(X_n \to Rx)$ to be known for all n, it is possible to determine $q(\theta)$ using the above equation. The particular case in which transmitters choose their destination uniformly among their N nearest neighbours will be covered in the next section.

For a given probability $q(\theta)$, which we denote simply as q, we have the following theorem.

Theorem 3: Conditioned on $\mathcal{I}_{Tx}(Rx) = k$, the probability for a data packet from a transmitter Tx to a receiver Rx and the following ACK packet from Rx to Tx to be both successfully received can be approximated by

$$p_{tx}^k\left(\theta,\theta'\,|\mathcal{I}_{Tx}(Rx)=k\right) \approx \Psi_2\left(\theta,\theta',1\right)^{k-1}$$

$$\frac{\mathbb{E}_{\mathcal{I}}\left\{\Psi_1\left(\theta,\theta',1\right)^{\mathcal{I}-1}\left[\frac{\Psi_3(\theta,\theta',1)}{\Psi_1(\theta,\theta',1)\Psi_2(\theta,\theta',1)}\right]^{\min(\mathcal{I},k)-1}_{\beta_d(1)}\right\}}{\left(1+\sum_{i=2}^{\infty}\Upsilon\left(\theta,\theta',i\right)\right)^k} \tag{19}$$

Where

$$\Psi_1\left(\theta,\theta'.i\right) = \left[\Phi\left(\theta'.i,i+1\right)\right]_q\left[\frac{\Phi\left(\theta,i-1,i\right)}{\left[\Phi\left(\theta',i,i+1\right)\right]_q}\right]_p$$

$$\Psi_2\left(\theta,\theta'.i\right) = \left[\Phi\left(\theta'.i-1,i\right)\right]_q\left[\frac{\Phi\left(\theta.i,i+1\right)}{\left[\Phi\left(\theta',i-1.i\right)\right]_q}\right]_p$$

$$\Psi_3\left(\theta,\theta'.i\right) = \left[\Phi\left(\theta'.i-1,i\right)\right]_q\left[\frac{\Phi\left(\theta.i-1,i\right)}{\left[\Phi\left(\theta',i-1.i\right)\right]_q}\right]_p$$

$$\Upsilon\left(\theta,\theta',i\right) = \left(i^d-\right)\left(2-\Psi_1\left(\theta,\theta',i\right)-\Psi_2\left(\theta,\theta',i\right)\right)$$

$$+\left(\beta_d(i)-2(i-1)^d+\beta_d(i-1)\right)\left(1-\Psi_3\left(\theta,\theta',i\right)\right) \tag{20}$$

and where $\beta_d(i) = \frac{|\mathcal{B}(Tx,ix)\cap\mathcal{B}(Rx,ix)|}{|\mathcal{B}(Tx,ix)|} = B_{\frac{1}{i}-\frac{1}{4i^2}}\left(\frac{d+1}{2},\frac{1}{2}\right).$

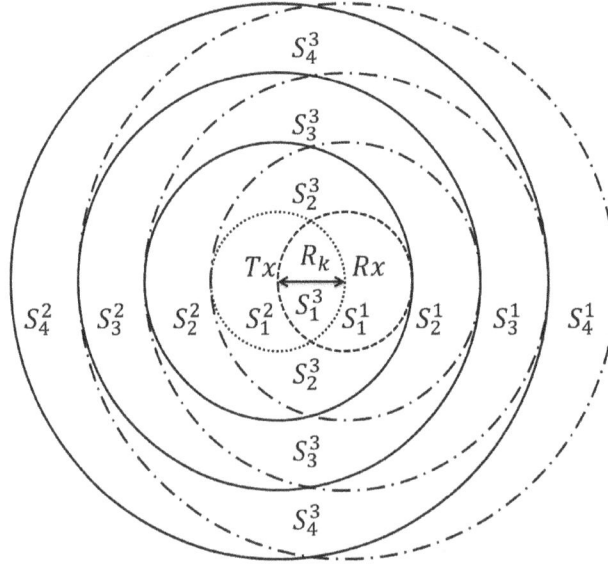

Figure 2 : Illustration of the space partition used for the analysis of transmission success probability using Slotted-ALOHA protocol with Acknowledgement packet. The partition is composed of the disjoint regions resulting from the intersections of two infinite sets of concentric circles, centred on Tx and Rx, respectively. The figure only shows the regions closest to Tx and Rx. Regions further away are defined in the same way.

Proof: Given that $\mathcal{I}_{Tx}(Rx) = k$, let us consider the partition $\left\{ \mathcal{S}_i^j(R) \right\}_{j=1,2,3;i=1,...,\infty}$ of $\mathbb{R}^d$, which is illustrated in figure 2. In this case the joint probability of success of a data transmission and its following ACK transmission is given by

$$p_{tx}^k(\theta,\theta') = \mathbb{E}\left\{ \prod_{j_1,i_1} L_{I_{\mathcal{S}_{i_1}^{j_1},Data}}(\theta R^\alpha) \prod_{j_2,i_2} L_{I_{\mathcal{S}_{i_2}^{j_2},Ack}}(\theta R^\alpha) \right\} \quad (21)$$

where the expectation is over R, $\mathcal{I}_{Rx}(Tx)$, $\left\{ \mathcal{N}_1\left(\mathcal{S}_i^j\right) \right\}_{j=1,2,3;i=1,...,\infty}$ and $\left\{ \mathcal{N}\left(\mathcal{S}_i^j\right) \right\}_{j=1,2,3;i=1,...,\infty}$. The first infinite product is actually the conditional probability of success of the data transmission, while the second product is that of the ACK transmission.

In [20], the authors employ product inequalities to separate the two transmit directions and derive lower and upper bounds. However, in our model, it is possible to treat directly the dependency between them. We can rewrite the handshake success probability using a single product as

$$p_{tx}^n(\theta,\theta') =$$
$$\mathbb{E}\left\{ \prod_{j,i} L_{I_{\mathcal{S}_i^j,Data}}(\theta R^\alpha) L_{I_{\mathcal{S}_i^j,Ack}}(\theta R^\alpha) \right\} \quad (22)$$

In order to keep the analysis tractable, we will use for each of the regions in the partition the same approximation on the distance distribution as we used to obtain (14). Note that according to the definition of q, the number of ACK packet transmitters in a given region $\mathcal{S}$ is binomially distributed according to $\mathfrak{B}\left(\mathcal{N}(\mathcal{S}) - \mathcal{N}_1(\mathcal{S}), q\right)$.

Consequently, we have

$$L_{I_{\mathcal{S}_i^1, Data}}(\theta R^\alpha) \approx \Phi(\theta, i-1, i)^{\mathcal{N}_1(\mathcal{S}_i^1)}$$

$$L_{I_{\mathcal{S}_i^2, Data}}(\theta R^\alpha) \approx \Phi(\theta, i, i+1)^{\mathcal{N}_1(\mathcal{S}_i^2)}$$

$$L_{I_{\mathcal{S}_i^3, Data}}(\theta R^\alpha) \approx \Phi(\theta, i-1, i)^{\mathcal{N}_1(\mathcal{S}_i^3)}$$

$$L_{I_{\mathcal{S}_i^1, Ack}}(\theta R^\alpha) \approx [\Phi(\theta, i, i+1)]_q^{\mathcal{N}(\mathcal{S}_i^1)-\mathcal{N}_1(\mathcal{S}_i^1)}$$

$$L_{I_{\mathcal{S}_i^2, Ack}}(\theta R^\alpha) \approx [\Phi(\theta, i-1, i)]_q^{\mathcal{N}(\mathcal{S}_i^2)-\mathcal{N}_1(\mathcal{S}_i^2)}$$

$$L_{I_{\mathcal{S}_i^3, Ack}}(\theta R^\alpha) \approx [\Phi(\theta, i-1, i)]_q^{\mathcal{N}(\mathcal{S}_i^3)-\mathcal{N}_1(\mathcal{S}_i^3)} \quad (23)$$

Also, $\mathcal{N}_1(\mathcal{S})$ is binomially distributed according to $\mathfrak{B}(\mathcal{N}(\mathcal{S}), p)$. Thus, replacing the above terms into (22) and taking the expectation over $\left\{\mathcal{N}_1\left(\mathcal{S}_i^j\right)\right\}_{j=1,2,3; i=1,...,\infty}$, we obtain

$$p_{tx}^k(\theta, \theta') \approx \mathbb{E}\left\{\prod_{i=1}^\infty \prod_{j=1}^3 \Psi_j(\theta, \theta', i)^{\mathcal{N}(\mathcal{S}_i^j)}\right\} \quad (24)$$

Regarding the terms $\mathcal{N}\left(\mathcal{S}_i^j\right)$, we observe that $\mathcal{N}(\mathcal{S}_1^3)$ is binomially distributed according to $\mathfrak{B}(\min(\mathcal{I}_{Rx}(Tx), k) - 1, \beta_d(1))$. Also, $\mathcal{N}(\mathcal{S}_1^2) = \mathcal{I}_{Rx}(Tx) - 1 - \mathcal{N}(\mathcal{S}_1^1)$ and $\mathcal{N}(\mathcal{S}_1^3) = k - 1 - \mathcal{N}(\mathcal{S}_1^1)$. Denoting $\mathcal{I} = \mathcal{I}_{Rx}(Tx)$, we then have

$$\mathbb{E}_{\mathcal{I}, \mathcal{N}(\mathcal{S}_1^1)}\left\{\prod_{j=1}^3 \Psi_j(\theta, \theta', 1)^{\mathcal{N}(\mathcal{S}_1^j)}\right\} = \Psi_2(\theta, \theta', 1)^{k-1}$$

$$\mathbb{E}_{\mathcal{I}}\left\{\Psi_1(\theta, \theta', 1)^{\mathcal{I}-1}\left[\frac{\Psi_3(\theta, \theta', 1)}{\Psi_2(\theta, \theta', 1)\Psi_2(\theta, \theta', 1)}\right]_{\beta_d(1)}^{\min(\mathcal{I}, k)-1}\right\} \quad (25)$$

The other terms $\left\{\mathcal{N}\left(\mathcal{S}_i^j\right)\right\}_{j=1,2,3; i>1}$ are independent of $\mathcal{I}_{Rx}(Tx)$, independent of one another and Poisson distributed with parameter $\lambda\left|\mathcal{S}_i^j\right|$.

Consequently, for $i > 1$ and any $j \in \{1, 2, 3\}$,

$$\mathbb{E}_{\mathcal{N}(\mathcal{S}_i^j)}\left\{\Psi_j(\theta, \theta', i)\right\} = e^{-\lambda\left|\mathcal{S}_i^j\right|(1-\Psi_j(\theta, \theta', i))} \quad (26)$$

Also, the area of the regions $\left|S_i^j\right|$ are given by

$$\left|S_i^1\right| = \left|S_i^2\right| = \left(i^d - \beta_d(i)\right)c_d R^d$$

$$\left|S_i^3\right| = \left(\beta_d(i) - 2(i-1)^d + \beta_d(i-1)\right)c_d R^d \quad (27)$$

Consequently, using the above and then taking the expectation with respect to R, we obtain an integral of the same type as the one leading to (14). Solving this integral leads directly to the final result, which completes the proof.

Given that most of the important results calculated until now are actually approximations, it is reasonable to compare the obtained formulae with the results of computer simulations.

5. Simulation Results

5.1. Simulation Method

The method we use in our computer program to simulate an infinite HPPP is that proposed in [12]. In each simulation drop, we generate a random deployment then record the success/failure of a given link. Results are finally averaged over all the deployment realizations.

The performance metric used is the Mean Maximum Achievable Spectral Efficiency (MMASE) η_{tx}^k conditioned on $\mathcal{I}_{Tx}(Rx) = k$, given by

$$\eta_{tx}^k\left(\theta, \theta'\right) = p\left(1-p\right)\log_2\left(1+\theta\right)p_{tx}^k\left(\theta, \theta'\right) \quad (28)$$

where the dependence on θ' disappears in the case no acknowledgement packet is used. In order to ease the computation of theoretical expressions, we use the following approximation for the function $\Omega(s, X)$.

$$\tilde{\Omega}(s, X) = \left(\left(1+sX^{-\alpha}\right)^{\frac{d}{\alpha}} - \left(sX^{-\alpha}\right)^{\frac{d}{\alpha}}\right)B\left(1-\frac{d}{\alpha}, 1+\frac{d}{\alpha}\right) \quad (29)$$

The above approximation is obtained easily from the limit property of the Hypergeometric function used in Appendix C.

5.2. Simulation results

In the following, we compare the theoretical results obtained in the previous sections with results from computer simulations. All theoretical results on figures are represented by solid lines, while dots are the simulated data. The path-loss exponent is $\alpha = 4$, $d = 2$ and $\lambda = 10^{-3}$ nodes per square-meter. On some figures (i.e when there was a noticeable difference), we plotted both theoretical results using the above approximation (solid lines) and using the expression with Hypergeometric function given in (4) (dashed lines).

Case of S-ALOHA without ACK: In figure 3 and figure 4, we consider the case of S-ALOHA in a uniform RAHN modelled as a 2-dimensional HPPP. We observe a good agreement between the behaviours of simulated and approximate theoretical results. The latter appear as upper bounds of the simulated ones, which is to be expected since the approximation used in the derivation of (20) overestimates the distance between some interferers and the receiver.

Figure 3 illustrates the trade-off between decodability and bit rate. Note that the optimal MCS for transmissions to a given neighbour depends on the operating MAP p. Conversely, for a given receiver threshold, intuition suggests the existence of an optimal MAP. This is confirmed in figure 4. Note that for p > 0.5, the number of receiver nodes is on average smaller than the number of transmitters, which implies that the optimal p lies between 0 and 0.5. We note that both theoretical and simulated results reach their respective maximum almost at the same abscissa, in both figures. Regarding the influence of θ on optimal MAP p, further investigations have showed that increasing the data packet receiver threshold θ decreases the optimal p, which follows directly from the trade-off between the density of transmissions and the individual transmission rate.

Case of S-ALOHA with ACK: Regarding S-ALOHA with acknowledgement, we assume a routing protocol which makes transmitters select their packet destination uniformly among their N closest neighbours. In this particular case, $q(\theta)$ is given as follows.

$$q\left(\theta\right) = \frac{p}{N} \sum_{k=1}^{N} p_{tx}^{k}\left(\theta\right) \quad (30)$$

where $p_{tx}^{k}\left(\theta\right)$ denotes the success probability for a transmission to the k-th nearest neighbour and is given by (19) where $\theta' = 0$. We assume $N = 2$ in the following. The performances of S-ALOHA with ACK packet are described in figure 5. The theoretical results are computed using (19) and (29). The denominator in (19) is approximated by computing the first thirty terms of the infinite sum (more terms did not provide any additional noticeable accuracy gain).

We observe again a good agreement between the behaviors of simulated and approximate theoretical results. The latter do not explicitly upper or lower bound simulated results, which can be explained by the diversity of shape of the regions on which the approximation for the distance from interferers to the receiver is used. The existence of optimal operating points can be noted, in a similar fashion to S-ALOHA. Again, both theoretical and simulated results reach their respective maximum at abscissa very close from each other.

Regarding the influence of θ' on optimal MAP p and θ, although it is not shown here due to the lack of space, increasing θ' also increases the optimal MAP p. This can be justified in the same way as the influence of θ on p in the case of S-ALOHA without acknowledgement, by considering that the number of interferers during the ACK packet slot is proportional to $1 - p$.
Also, by comparing figure 5 and figure 6, we observe that increasing θ' actually increases the value of the optimal θ, although the achieved maximum MMASE is lower than with a smaller θ'. One possible justification for this result is that a higher θ leads to less interference during the ACK slot, thus allowing a higher θ' to be used.

Finally, comparing Fig 3, Fig 5 and Fig 6, we observe that the transmission of the ACK packet has negligible influence for the case $k = 1$, even for high values of θ'. Note however, that for $k > 1$, the influence of imperfect feedback becomes stronger (e.g for $k = 3, \theta' = 0\text{dB}$, the MMASE in the case of S-ALOHA with ACK is 75% of the case without ACK). Our observations confirm the intuitive fact that imperfect feedback is only negligible for transmissions between close neighbors.

6. Conclusion

In this paper, we analyzed the probability of success of a transmission (one-way and handshake) to the k-th nearest neighbor in an ad hoc network modelled by a single homogeneous Poisson Point Process in which nodes employ the Slotted-ALOHA MAC protocol (without and with acknowledgement, respectively). From the principle of maximum entropy, the single HPPP is the best deployment model for ad hoc networks with a constant node density. Previous works so far have been limited to a simplified deployment model, for tractability reasons. However, in this work, we presented a tractable mathematical analysis enabling the analysis of spatially averaged network performances at the routing level while tackling directly the dependency between both transmit directions in handshakes. We compared the developed close-form formula with computer simulation results and concluded that this work can find applications in the joint quantitative study of some MAC and routing protocols. Notably, exploiting the notion of neighbor index, we confirmed through our results the intuitive fact that imperfect feedback in handshakes is only negligible for transmissions to/from the closest neighbor.

Appendix A

In this appendix, we provide the proof of (3). From [13] (eq.(4)), we have

$$L_I(s) = \left(\frac{d}{B^d - A^d} \int_A^B x^{d-1} L_G(sx^{-\alpha}) \, dx \right)^k \quad (31)$$

The Laplace transform of the squared fading is given by

$$L_G(s) = \frac{1}{1+s} \quad (32)$$

Replacing the above in (31), leads to the integral given by

$$J = \int_A^B \frac{x^{d-1} dx}{1 + sx^{-\alpha}} = \int_A^B u(x) \, dx = \int_0^B u(x) \, dx - \int_0^A u(x) \, dx \quad (33)$$

Note that through the variable substitution $t = 1 - \left(\frac{x}{y} \right)^\alpha$, each of the integrals in the last expression of (34) can also be expressed as

$$\int_0^y u(x) \, dx = \frac{y^d \int_0^1 (1-t)^{\frac{d}{\alpha}} \left(1 - \frac{1}{1+sy^{-\alpha}} t \right)^{-1} dt}{\alpha (1 + sy^{-\alpha})} \quad (34)$$

where y takes the value A or B. One can recognize in (34) Euler's integral

transform for the Gauss hypergeometric function[17], given by (35).

$$F(a, b, c, z) = \frac{\int_0^1 t^{b-1} (1-t)^{c-b-1} (1-zt)^{-a} \, dt}{B(b, c-b)} \quad (35)$$

where $B(a, b)$ is the Beta function and $F(a, b; c; z)$ denotes the Gauss hypergeometric function. It then follows that

$$\int_0^y u(x) \, dx = \frac{y^d F\left(1, 1, 2 + \frac{d}{\alpha}, \frac{1}{1+sy^{-\alpha}}\right)}{(\alpha + d)(1 + sy^{-\alpha})} \quad (36)$$

Then, one can use the following identity for $z = \frac{sx^{-\alpha}}{1+sx^{-\alpha}}$ to relate the above result to the function $\Omega(s, X)$ as defined earlier in the paper

$$\frac{F\left(1, 1, 2 + \frac{d}{\alpha}, 1 - z\right)(1-z)\frac{d}{\alpha}}{1 + \frac{d}{\alpha}} = F\left(1, 1, 1 - \frac{d}{\alpha}, z\right)(1-z) - B\left(1 - \frac{d}{\alpha}, 1 + \frac{d}{\alpha}\right)\left(\frac{z}{1-z}\right)^{\frac{d}{\alpha}}$$

$$(37)$$

Using (37) and (36) in (31) leads to the final result, which completes the proof.

Appendix B

In this appendix we provide the proof of (5). We first notice from the properties of the Gauss Hypergeometric function that for $c - a - b < 0$, we have [17]

$$\lim_{z \to 1^-} \frac{F(a,b,c,z)}{(1-z)^{c-a-b}} = \frac{\Gamma(c)\Gamma(a+b-c)}{\Gamma(a)\Gamma(b)} \qquad (38)$$

Then, using (4), let us rephrase $A^d \Omega(s,A)$ as

$$A^d \Omega(s,A) = A^d f(s,A)\left(1 - \frac{s}{s+1A^\alpha}\right)^{-\frac{d}{\alpha}} - A^d - B\left(1-\frac{d}{\alpha}, 1+\frac{d}{\alpha}\right)s^{\frac{d}{\alpha}} \qquad (39)$$

Where $f(s,A) = F\left(1,1,1-\frac{d}{\alpha}, \frac{s}{1A^\alpha+s}\right)\left(1 - \frac{s}{s+1A^\alpha}\right)^{1+\frac{d}{\alpha}}$. We observe that by setting $z = \frac{s}{s+1A^\alpha}$ and $a=1, b=1, c=1-\frac{d}{\alpha}$, we obtain easily the limit of $f(s,A)$ for A converging toward zero from (38). From there, the final result easily follows, which completes the proof.

Appendix C

In this appendix we provide the proof of (6). From [12] (section 3.7.1), we have

$$L_{I_{\mathbb{R}^d/\mathcal{B}(O,x)}}(s) = e^{-\lambda c_d \mathbb{E}_G\left\{s^{\frac{d}{\alpha}}G^{\frac{d}{\alpha}}\gamma\left(1-\frac{d}{\alpha}, sGx^{-\alpha}\right) - x^d\left(1 - e^{-sGx^{-\alpha}}\right)\right\}} \qquad (40)$$

where $\gamma(n,x)$ denotes the lower incomplete gamma function. Let us rephrase part of the above expression as follows.

$$\mathbb{E}_G\left\{G^{\frac{d}{\alpha}}\gamma\left(1-\frac{d}{\alpha}, sGx^{-\alpha}\right)\right\} = \int_0^{+\infty} \frac{1}{\Gamma(1)} t^{\frac{d}{\alpha}}\gamma\left(1-\frac{d}{\alpha}, tsx^{-\alpha}\right)e^{-1t}dt \qquad (41)$$

The above integral can be solved by replacing initially the incomplete lower gamma function by its equivalent involving the Kummer's confluent hypergeometric function.[17] That is,

$$\gamma\left(1-\frac{d}{\alpha}, tsx^{-\alpha}\right) = \frac{e^{-tsx^{-\alpha}}M\left(1, 2-\frac{d}{\alpha}, tsx^{-\alpha}\right)}{\left(1-\frac{d}{\alpha}\right)\left(tsx^{-\alpha}\right)^{\frac{d}{\alpha}-1}} \qquad (42)$$

Then, replacing (42) into (41) reveals an integral form of the Gauss Hypergeometric function given by

$$F(a,b,c,z) = \frac{1}{\Gamma(b)}\int_0^{+\infty} e^{-t}t^{b-1}M(a,c,tz)dt \qquad (43)$$

which is valid only for $b > 0$. Therefore, it follows that

$$\mathbb{E}_G\left\{G^{\frac{d}{\alpha}}\gamma\left(1-\frac{d}{\alpha}, sGx^{-\alpha}\right)\right\} = \frac{(sx^{-\alpha})^{1-\frac{d}{\alpha}}F\left(1,2,2-\frac{d}{\alpha}, \frac{sx^{-\alpha}}{1+sx^{-\alpha}}\right)}{\left(1-\frac{d}{\alpha}\right)(1+sx^{-\alpha})^2}$$

$$= \frac{zs^{-\frac{d}{\alpha}}x^d}{\left(1-\frac{d}{\alpha}\right)}F\left(1,2,2-\frac{d}{\alpha}, z\right)(1-z) \qquad (44)$$

Where $z = \frac{sx^{-\alpha}}{1+sx^{-\alpha}}$. Note that using the same notation, we have

$$\mathbb{E}_G\left\{1 - e^{-sGx^{-\alpha}}\right\} = 1 - (1-z) \qquad (45)$$

Using the recurrence relations of the Hypergeometric function, one can obtain the following identity

$$F(1,b,c,t) = \frac{c-1}{b-1}\left(\frac{F(1,b-1,c-1,t)}{t} - \frac{1}{t}\right) \qquad (46)$$

which, by setting $b = 2$, $c = 2 - \frac{d}{\alpha}$ and $t = z = \frac{sx^{-\alpha}}{1+sx^{-\alpha}}$ lead us to

$$
\mathbb{E}_G \left\{ G^{\frac{d}{\alpha}} \gamma \left(1 - \frac{d}{\alpha}, sGx^{-\alpha} \right) \right\} = s^{-\frac{d}{\alpha}} x^d \left(F \left(1, 1, 1 - \frac{d}{\alpha}, z \right) - 1 \right) (1 - z) \quad (47)
$$

Replacing (47) and (45) into (40) we obtain

$$
L_{I_{\mathbb{R}^d/\mathcal{B}(O,x)}} (s) = \exp \left(-\lambda c_d x^d \left(F \left(1, 1, 1 - \frac{d}{\alpha}, z \right) (1 - z) - 1 \right) \right) \quad (48)
$$

which can be expressed as a function of $\Omega(s, x)$ as defined in (4) and thus completes the proof.

Appendix D

In this appendix, we prove that the p.m.f in (8) satisfies the normalization condition. That is, we aim to prove that

$$
S_k = \sum_{n=1}^{\infty} \mathcal{F}(n, k) = 1 \quad (49)
$$

To do so, we start by reordering the double sum above into

$$
S_k = \sum_{l=0}^{k-1} \sum_{n=l+1}^{\infty} \binom{k-1}{l} \binom{n+k-l-2}{k-1} \frac{\beta_d^l (1 - \beta_d)^{n+k-2l-2}}{(2 - \beta_d)^{n+k-l-1}} \quad (50)
$$

which is obtained simply by interverting sum indexes after making the term $\min(n, k)$ disappear. Then, sorting terms and using the change of index $n = n' + l + 1$, we obtain

$$
S_k = \frac{1}{1 - \beta_d} \left(\frac{1 - \beta_d}{2 - \beta_d} \right)^k \sum_{l=0}^{k-1} \binom{k-1}{l} \left(\frac{\beta_d}{1 - \beta_d} \right)^l \sum_{n=0}^{\infty} \binom{n+k-1}{n} \left(\frac{1 - \beta_d}{2 - \beta_d} \right)^n
$$
(51)

Using the binomial theorem and recognizing the series expansion of the function $(1 - x)^{-k}$ in the second sum, we can transform the above into

$$
S_k = \frac{1}{1 - \beta_d} \left(\frac{1 - \beta_d}{2 - \beta_d} \right)^k \frac{1}{(1 - \beta_d)^{k-1}} (2 - \beta_d)^k = 1 \quad (52)
$$

which completes the proof, for all $k \geq 1$.

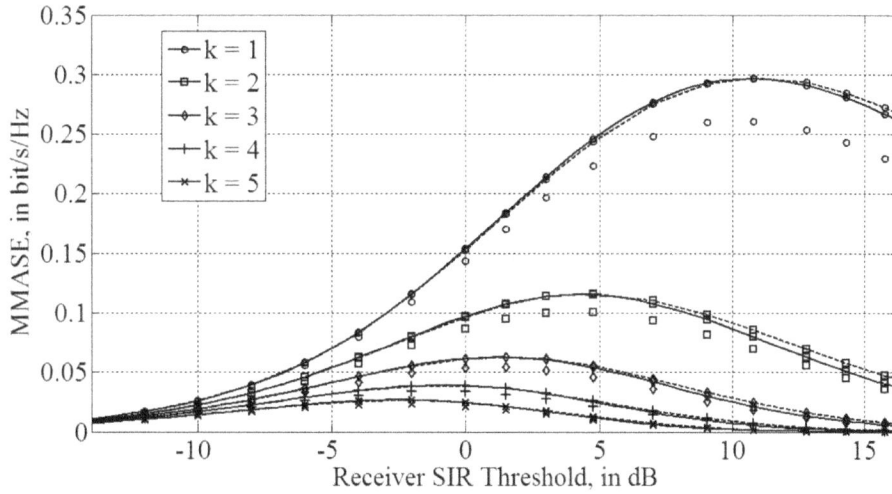

Figure 3 : MMASE of transmissions to the k-th neighbour in a 2-dimensional uniform RAHN employing S-ALOHA with fixed MAP $p = 0.3$, against the receiver threshold (in dB).

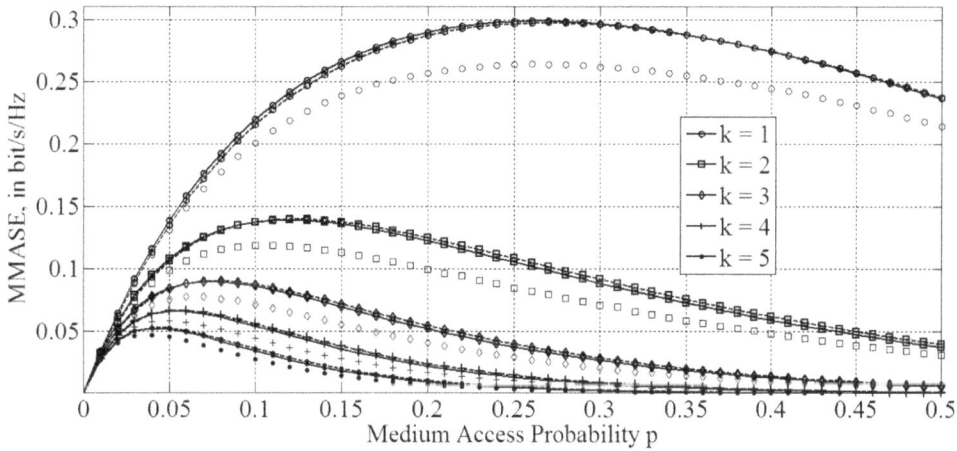

Figure 4 : MMASE of transmissions to the k-th neighbour in a 2-dimensional uniform RAHN employing S-ALOHA, against the MAP p, for a given a receiver threshold $\theta = 10.4$.

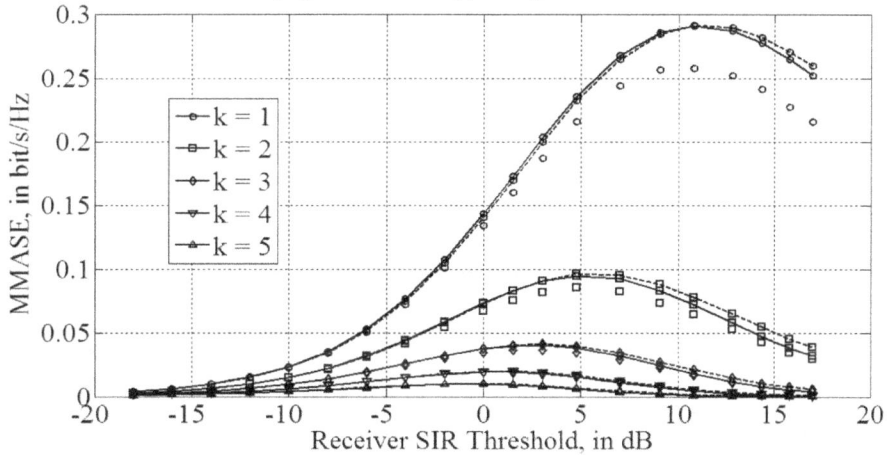

Figure 5 : MMASE of transmissions to the k-th neighbour in a 2-dimensional uniform RAHN employing S-ALOHA with ACK and fixed MAP $p = 0.3$, against the receiver threshold θ (in dB). The curves are obtained from (19), for $\theta' = 0dB$.

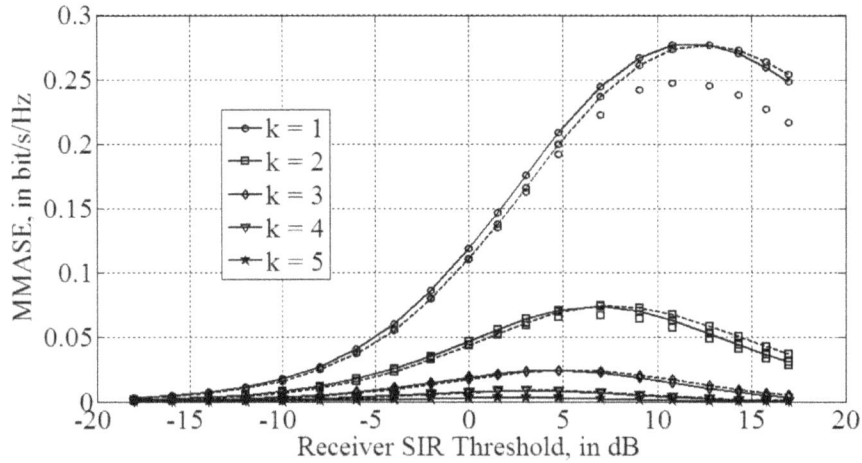

Figure 6 : MMASE of transmissions to the k-th neighbour in a 2-dimensional uniform RAHN employing S-ALOHA with ACK and fixed MAP $p = 0.3$, against the receiver threshold θ (in dB). The curves are obtained from (19), for $\theta' = 9dB$.

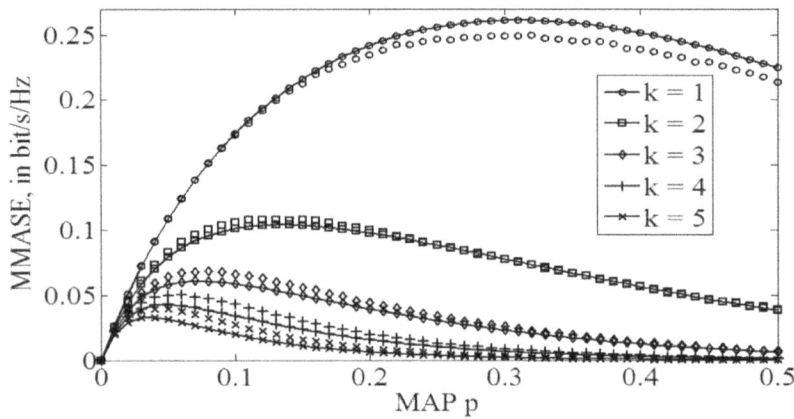

Figure 7 : MMASE of transmissions to the k-th neighbour in a 2-dimensional uniform RAHN employing S-ALOHA with ACK packet, against the MAP p. The data packet threshold is $\theta = 8dB$ and the ACK packet threshold is $\theta' = 0dB$. The curves are obtained from (19).

REFERENCES

[1]   J. Akhtman and L. Hanzo. Heterogeneous networking: An enabling paradigm for ubiquitous wireless communications [point of view]. Proceedings of the IEEE, 98(2):135–138, Feb.

[2]   .G. Andrews, R.K. Ganti, M. Haenggi, N. Jindal, and S. Weber. A primer on spatial modeling and analysis in wireless networks. Communications Magazine, IEEE, 48(11):156 –163, november 2010.

[3]   Weng Chon Ao and Kwang-Cheng Chen. Bounds and exact mean node degree and node isolation probability in interference-limited wireless ad hoc networks with general fading. Vehicular Technology, IEEE Transactions on, 61(5):2342–2348, 2012.

[4]   F. Baccelli, B. Blaszczyszyn, and P. Muhlethaler. An aloha protocol for multihop mobile wireless networks. Information Theory, IEEE Transactions on, 52(2):421 – 436, feb. 2006.

[5]   F. Baccelli, Bartlomiej Blaszczyszyn, and P. Muhlethaler. Stochastic analysis of spatial and opportunistic aloha. Selected Areas in Communications, IEEE Journal on, 27(7):1105–1119, 2009.

[6]   S. Basagni, M. Conti, S. Giordano, and I. Stojmenovic. Mobile Ad Hoc Networking. Wiley-IEEE Press, 2004.

[7]   T. Bourgeois and S. Shimamoto. Stochastic analysis of handshake-type mechanisms in uniformly random wireless ad hoc networks. submitted to Global Telecommunications Conference, 2013. GLOBECOM '13. IEEE.

[8]   D.J. Daley and D. Vere-Jones. An Introduction to the Theory of Point Processes. Springer Series in Statistics, 1988.

[9] R.K. Ganti, J.G. Andrews, and M. Haenggi. High-sir transmission capacity of wireless networks with general fading and node distribution. Information Theory, IEEE Transactions on, 57(5):3100–3116, 2011.

[10] M. Haenggi. A geometric interpretation of fading in wireless networks:Theory and applications. Information Theory, IEEE Transactions on, 54(12):5500–5510, 2008.

[11] M. Haenggi. Outage, local throughput, and capacity of random wireless networks. Wireless Communications, IEEE Transactions on, 8(8):4350–4359, 2009.

[12] M. Haenggi and R. K. Ganti. Interference in large wireless networks. Foundations and Trends in Networking, (2):127 –248, 2009.

[13] M. Haenggi and S. Srinivasa. Modeling interference in finite uniformly random networks. Proc. International Workshop on Information Theory for Sensor Networks (WITS'07), 2007.

[14] H. Inaltekin, Mung Chiang, H.V. Poor, and S.B. Wicker. On unbounded path-loss models: effects of singularity on wireless network performance. Selected Areas in Communications, IEEE Journal on, 27(7):1078–1092, September.

[15] E. T. Jaynes. Information theory and statistical mechanics. Physical Review, 106:620 – 630, May 1957.

[16] E. T. Jaynes. Information theory and statistical mechanics. Physical Review, 108:171 – 191, Oct. 1957.

[17] F. W. J. Olver, D. W. Lozier, R. F. Boisvert, and C. W. Clark, editors. NIST Handbook of Mathematical Functions. Cambridge University Press, New York, NY, 2010.

[18] S.P. Smith. Threshold validity for mutual neighborhood clustering. Pattern Analysis and Machine Intelligence, IEEE Transactions on, 15(1):89 –92, jan 1993.

[19] R. Vaze and R.W. Heath. Transmission capacity of ad-hoc networks with multiple antennas using transmit stream adaptation and interference cancellation. Information Theory, IEEE Transactions on, 58(2):780–792, 2012.

[20] R. Vaze, K.T. Truong, S. Weber, and R.W. Heath. Two-way transmission capacity of wireless ad-hoc networks. Wireless Communications, IEEE Transactions on, 10(6):1966–1975, 2011.

[21] J. Zheng and A. Jamalipour. Wireless Sensor Networks:A Networking Perspective. Wiley-IEEE Press, 2009.

[22] E. Chukwuka and K. Arshad. Energy Efficient MAC Protocols for Wireless Sensor Network: A Survey. International Journal of Wireless and Mobile Networks (IJWMN), 5(4), August 2013

[23] V. Hnatyshin. Improving MANET Routing Protocols Through the Use of Geographical Information. A Survey. International Journal of Wireless and Mobile Networks (IJWMN), 5(2), April 2013

6

Impact of Macrocellular Network Densification on the Capacity, Energy and Cost Efficiency in Dense Urban Environment

S. F. Yunas, T. Isotalo, J. Niemelä and M. Valkama

Department of Electronics and Communications Engineering,
Tampere University of Technology, Tampere, Finland

ABSTRACT

This paper aims to show the effect of macrocellular network densification on the capacity, energy and cost efficiency. The presented results are based on radio propagation simulations that consider macrocellular network with different inter-site distances, i.e. different site densities, and also take into account the presence of indoor receiver points by varying outdoor and indoor receiver distribution. It is observed that as a result of densifying the network, the cell spectral efficiency reduces due to increasing level of inter-cell interference. However, as a result of densification, the network area capacity can be improved since the area spectral efficiency increases. Nevertheless, the densification efficiency decreases because of the reduction of cell spectral efficiency, especially when indoor receiver points are taken into account. The results hence indicate that densification of macrocellular network suffers from inefficiency which results in higher energy and cost per bit per Hertz, and thus calls for alternative methods to deploy networks, or alternatively, more sophisticated methods, such as base station coordination or inter-cell interference cancellation techniques, to be implemented for future cellular networks.

KEYWORDS

Macrocell densification, area spectral efficiency, energy efficiency, cost efficiency, 3D ray tracing model

1. INTRODUCTION

Macrocellular networks have been and still continues to be the basis for cellular network deployments globally. High power transmitters with highly elevated and directive antenna array positions are superior in terms of wide-area coverage provisioning. They also play a major role in fulfilling the mobility demands of cellular users, and hence, are assumed to maintain their position in the future as well. Moreover, it is envisioned that macrocell networks will continue to provide the outdoor coverage layer with small cells satisfying the local outdoor and indoor capacity demands. Current cellular networks are inherently heterogeneous in terms of network configuration and this trend is building towards an even denser heterogeneous configuration as new small cell technology and different other indoor network solutions become more and more common.

The global cellular traffic has been on a steady rise since the early days of 2G networks. With the increasing popularity of mobile broadband subscriptions, the projected growth in the capacity demand has been estimated to increase 15 folds in the upcoming five years [1]. Fundamentally, the main mechanisms to increase the network capacity are increased link and radio resource management efficiency together with utilization of wider bandwidth and the cell size, i.e., having dense network configuration with small cells. The idea of enhancing the system capacity through network densification can be dated back to late 1940s when the cellular concept was introduced [2]. The initial adoption of the concept, however, was slow at first but started to gain serious attention when 2G networks were introduced. Since then, network densification has been viewed as a feasible pathway towards network evolution.

Lately, engineers and academic researchers have been studying and evaluating the concept with different performance enhancing techniques. In [3], the performance of macrocellular densification with different transmission schemes has been compared with a network employing base station coordination algorithms. The study concentrates on various techniques that can maximize the minimum spectral efficiency of the served users. In addition, a constant user density, irrespective of the network size, has been assumed. This results in a partially loaded system where some of the base stations are kept in sleep mode to avoid over provisioning of the network capacity. The results show that the cell spectral efficiency increases as the network is densified to a certain point and then saturates. In [4], the average cell spectral efficiency is shown to increase linearly with network densification in partially loaded system. The impact of macrocell densification on the network throughput and power consumption in both homogeneous and heterogeneous network environments has been studied in [5]. The study considers a fully loaded network, where all the base stations are continuously transmitting at full power. However, the maximum transmit power per base station is varied as the network is densified. The findings in [5] follow the outcomes of [4] i.e., in a homogeneous macrocell network, the cell spectral efficiency tends to improve with increasing network density. In [6, 7], the performance of homogeneous macrocellular network densification has been examined and compared with different heterogeneous network deployment alternatives. The papers take a slightly different approach by introducing variable traffic. Hence the system performance is evaluated in terms of served area traffic during busy hour. Unlike in [3-5], where only an outdoor environment is assumed, the studies in [6, 7], also take into account the indoor environment with buildings and users distributed among different floors. Nevertheless, their results indicate increasing served area traffic per busy hour as the network is densified.

While significant amount of time and effort in the last two decades was dedicated by the industry and academia in improving the spectral efficiency of wireless networks, more recently, the focal point of the industry has started to expand towards including energy and cost efficiency aspects into its domain. To cope with the current rate of 'exponentially' increasing capacity demand, deployment of several magnitudes more base stations will be required, which is considered by the industry to be a feasible pathway. However, this strategy is known to significantly increase the cost and energy consumption of the cellular networks. The ICT industry currently contributes around 2% of the global greenhouse gas emission and this contribution is expected to increase to 4% by the year 2020 [8]. According to some studies conducted in 2007/2008, the radio access networks alone had a share of around 0.3% - 0.5% in the global CO_2 emissions [9, 10] and out of this roughly 80% came from the base stations [11]. As the worldwide awareness regarding global warming increases, political initiatives at the international level have started to put requirements on the operators and manufacturers to lower the gas emissions of communications networks [12]. This has led the telecommunication industry, especially the standardization and regulatory bodies, to focus their attention towards building 'greener' wireless networks. In the research community, considerable number of studies have been conducted and published in the recent years focusing on quantifying the energy consumption of the wireless networks by establishing different metrics for evaluation of the energy efficiency, proposing power consumption models for different base station types and ways to improving the power consumption of the networks while maintaining decent quality of service and system throughput. Studies emphasizing on the importance of having a holistic framework for evaluating the energy efficiency of the wireless networks have been reported in [13, 14]. In [13], the authors discuss the importance of evaluating the energy efficiency at each level of network hierarchy, namely component, link and network level, their mutual dependencies and the need for optimizing the system as a whole, rather than just focusing on improving only one aspect, for achieving an optimum system performance. Building up on the foundation laid in [13], the authors propose an energy efficiency evaluation framework in [14] for evaluating nationwide energy-efficiency of a mobile operator. The framework constitutes a power consumption model for different LTE base station types together with a proposed large scale deployment and long term traffic model. In [11], a new metric, *area power consumption*, is proposed to evaluate and compare the energy efficiencies of networks with different cell ranges (varying cell site densities/km^2). A brief overview of the energy efficiency metrics at different levels of a wireless network/system has been provided in [15]. The impact of cell size on the power consumption has been studied in [4] and [16-18] for different deployment strategies. However, the results in [16] differ with the other studies. The authors in [16] report that large cell deployments are efficient in terms of area power consumption as compared to small cells while the finding of [4], [17] and [18] claim otherwise. The contradictory results reported in the above studies come from differences in the power consumption models utilized in the studies and using different energy efficiency metrics that capture different statistics of the network [19]. The impact of network densification on the energy efficiency in the wireless networks has been investigated in [19]. Unlike the previous studies in [4] and [16-18], which fail to take into account the impact of interference and system throughput while evaluating the energy efficiency, the studies in [19] investigate the relation between energy efficiency, area capacity and cell size by taking into consideration both the interference and noise, and takes relates the energy efficiency in terms of system throughput. The paper refines the analytical power consumption model proposed in [11] to include the backhaul power consumption as well which has significant impact when considering dense deployments [20]. The studies reported in [15-19] have a common short come, i.e. they all use a simple analytical model that combines the power consumption of components into two or three parameters and hence do not accurately model the influence of each component, within the base station, on the total power consumption. In [14], the power consumption of core components that have significant impact on the total power consumption of the base station, have been separately modelled. Nevertheless, it still fails to take into account the backhaul power consumption in its model, which is a key in denser deployments, as mentioned in [20]. This problem has been addressed in [21], where the proposed analytical model takes into account the contribution of the core components within the base station as well as the influence of backhaul transmission unit. The performance of the proposed power consumption model is further validated by comparing it with measurements from a live 3G network.

In this paper, we study the downlink capacity performance of macrocellular network densification in a full load condition, which is the worst case scenario and also a typical methodology that is used for network capacity dimensioning. Unlike in [4] and [5], where the base station parameters (antenna height, transmit power) are varied with respect to the cell size, we keep the antenna height and transmit power unchanged; only the downtilt angle is varied as network is densified. Further, an accurate deterministic 3D ray tracing model has been used in our study. For energy-efficiency analysis we use the power consumption model given in [21]. The main target of the paper is to evaluate impact of site densification on the average cell and area spectral efficiency for outdoor and indoor receiver point locations, and also to evaluate whether macrocellular densification is a feasible pathway, in terms of energy and cost efficiency, towards the evolution of future cellular networks. The study aims to answer the question, *how much system capacity gain can we achieve through macrocellular network densification and whether the capacity gain is enough lower the energy per bit and cost per bit to make pure macrocellular densification energy and cost efficient solution?* The rest of the paper is structured as follows. We introduce the system model in Section 2 with the analysis methodology. Section 3 presents the results and performance analysis for capacity efficiency. In Section 4, the power consumption model for the base stations is introduced and the energy efficiency results and analyzed. Next, the cost modelling methodology is discussed in Section 5 and the cost efficiency results are analyzed. Finally, based on the analysis, concluding remarks are given in Section 6.

2. SYSTEM MODELLING

This section presents the system model that we have used in our macrocellular densification studies. It starts with the introduction of the simulation environment and cell layouts. This is followed by a brief description of the simulation tool and the propagation model that was used in evaluations. Next, different network and site configuration parameters as antenna model together with average inter-site distance (ISD) and corresponding tilt angles are described. Finally, a list of general simulation parameters is provided.

2.1. Simulation environment and cell layout

To imitate a dense urban environment, we created a fictive Manhattan type grid city model (Figure 1). A hexagonal layout was used as the basis of our macrocellular deployment strategy, although the actual network layout determined by the Manhattan grid that does not allow a pure hexagonal cell layout. Each building has dimensions of 110 m x 110 m, a height of 40 m and comprises of 8 floors. The streets were selected to be 30 m wide. For indoor floor plan, an open office layout was chosen. It renders a hall area with no rooms, i.e. no hard obstruction for signal propagation except for the ceiling, floors and exterior walls.

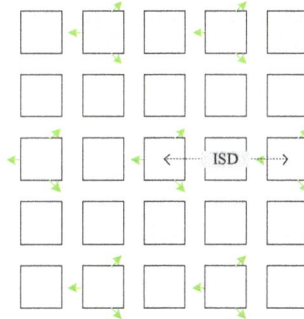

Figure 1. Manhattan grid city model (aerial view). The arrows show sector antenna positions and orientations (example case: ISD 297 m). Note that only the first interfering tier is illustrated. The appropriate number of interfering tiers for each configuration used in the simulations is given in Table 2.

2.2. Propagation model

A commercial radio wave propagation tool (Wireless Insite), was used for the coverage prediction simulations. The outdoor and indoor radio channels are modelled using a 3D ray based model. The model employs a ray-launching technique based on 'Shooting and Bouncing Ray' (SBR) method to find the propagation paths through the 3D building geometry between a transmitter and receiver [22]. Rays are shot from the emitting source in discrete intervals and traced correspondingly as they reflect, diffract and transmit (penetrate) through and around the obstacles. Each ray is traced independently and the tracing continues until the maximum number of reflections is reached. Once all the propagation paths have been computed and stored, the field strength for each ray path is calculated using Uniform Theory of Diffraction (UTD) [23-25].

The accuracy of a 3D propagation model is dependent upon the input data and the total number of reflections, transmissions (or wall penetrations) and diffractions a single ray can encounter. To limit the calculation time, we used an empirical 'hit-and-trial' method, which involves simulating with a smaller number of interactions, and then re-simulating the same scenario by steadily increasing interactions and comparing the results. Once the results start to converge with insignificant change, those settings were then selected. In our case, this was observed at 10 reflections, 1 diffraction and 1 transmission where providing sufficiently accurate results.

To account for the outdoor-to-indoor propagation loss, the exterior wall direct penetration loss was chosen to be 25 dB. The corresponding electrical properties of the exterior wall were found empirically by adjusting the conductivity σ, permittivity ε and thickness of the wall, and observing the difference in the average signal level between several outdoor and indoor receiver points.

2.3. Antenna models and positions

An extended 3GPP antenna model based on [26] was adopted for simulations. The proposed version extends the original model of [13], which only considers the horizontal plane, and includes a vertical antenna pattern model with an option to set the electrical downtilt. The horizontal (azimuth) pattern, G_h, is given by:

$$G_h(\varphi) = -\min\left[12\left(\frac{\varphi}{HPBW_h}\right)^2, FBR_h\right] + G_m \tag{1}$$

where, φ, $-180^0 \leq \varphi \leq 180^0$, is the azimuth angle relative to the main beam direction, $HPBW_h$ is the horizontal half power beam width [^{0}], FBR_h is the front-to-back ratio [dB] and G_m is the maximum gain of the antenna [dBi]. The vertical (elevation) pattern, G_v, is given by:

$$G_v(\phi) = -\max\left[-12\left(\frac{\phi - \phi_{etilt}}{HPBW_v}\right)^2, SLL_v\right] \tag{2}$$

where ϕ, $-180^0 \leq \phi \leq 180^0$, is the negative elevation angle relative to horizontal plane (i.e., $\phi = -90^0$ is the upward plane relative to the main beam, $\phi = 0^0$ is along the main beam direction, and $\phi = 90^0$ is the downward plane relative to the main beam), ϕ_{etilt} is the electrical downtilt angle [^{0}], $HPBW_v$ is the vertical half power beam width [^{0}], and SLL_v is the side lobe level [^{0}] relative to the maximum gain. The antenna parameter values were adopted from [26] except for electrical tilt angles which were based on the average inter-site distances.

The antennas were placed 2 m above the building roof, i.e. 42 m above the ground level. In order to ensure that the transmitted signal is not obstructed by the roof, sector antennas were placed either at corners of the buildings rather than at the center to ensure unobstructed propagation (see arrows in Figure 1).

2.4. ISDs (cell density) and electrical tilt

The cell density depends upon the average inter site distance (ISD or $\bar{d}_{site}$), which further specifies the dominance area of a cell. In our study, we define dominance area as the region where a cell provides highest signal level as compared to the rest of the cells. Altogether five different ISDs were considered. These were calculated from the center of the building (except in the average ISD of 170 m case, where it is calculated based on average inter cell distance owing to the square layout of the buildings).

Assuming a regular hexagon cell, the dominance area of a cell, A_{cell} is given by:

$$A_{cell}\left[km^2\right] = \frac{\sqrt{3}}{6}\left(\bar{d}_{site}\right)^2 \tag{3}$$

The cell density, ρ_{cell}, per km^2 is defined as $1/A_{cell}$.

In order to avoid unnecessary interference into neighboring cells, the sectors are required to be down tilted. As mentioned earlier, the electrical tilt angle depends on the ISD as it defines the maximum cell range. Knowing the base station (BS) antenna height (h_{BS}), the mobile station (MS) antenna height (h_{MS}) and the cell range (r_{cell}), the tilt angle was calculated geometrically as:

$$\phi_{etilt} = \arctan\left(\frac{h_{BS} - h_{MS}}{r_{cell}}\right) \qquad (4)$$

The rest of the simulation parameters are gathered in Table 1. Note that the effective isotropic radiated power (EIRP) in the maximum antenna gain direction is 43 dBm + 18 dBm = 61 dBm. Moreover, for the receiver noise floor level calculation, a 20 MHz bandwidth was assumed (nominal for long term evolution, LTE).

Table 1. General simulation parameters.

Parameter	Unit	Value
Operating frequency	[MHz]	2100
Bandwidth, W	[MHz]	20
Transmit power at the antenna, P_{TX}	[dBm]	43
BS antenna beam width, $HPBW_{h/v}$	[degrees]	Directional ($65^0/6^0$)
MS antenna type		Half-wave dipole
BS antenna gain	[dBi]	18
MS antenna gain	[dBi]	2
BS antenna height, h_{BS}	[m]	42
MS antenna height, h_{MS}	[m]	2
Receiver noise figure	[dB]	9
Receiver noise floor, P_n	[dBm]	-92
Propagation environment		Manhattan
Propagation model		3D ray tracing
Building dimensions	[m]	110 x 110
Building height	[m]	40
Street width	[m]	30
Indoor layout		Open office
Outdoor-to-indoor wall penetration loss	[dB]	25

2.4. Analysis methodology

Due to homogeneity of the environment, we only consider the receiver points from the dominance area of the center cell site for statistical analysis and then normalize the analysis to 1 km^2 area. For simulating a continuous cellular network and analyzing its performance, it is necessary to take into account all the interfering cells that contribute to the interference level in the dominance area of a serving cell. In a realistic environment, the transmission path loss actually caps the number of receivable interfering sources at the serving cell which in turn limits the total interference level. However, in an ideal environment, like the Manhattan grid, the situation is exacerbated by the street canyons. The signals travelling in a street canyon tend to travel further as compared to the signal travelling in free space [27]. This is due to the tunneling effect caused by the side walls which direct the signals into the alley. As a consequence, the effect of distant interfering tiers especially the LOS (line of sight) tiers, which were negligible before, starts to become more visible.

To estimate the number of interfering tiers, only those tiers that had significant impact on the relative interference levels at the serving cell border (which is worst case scenario) are considered. This hit and trial method provides a fast and reliable estimate of the effective/dominant interfering tiers. Table 2 lists the average inter-site distance ($\overline{d}_{site}$) and the corresponding electrical tilt angles, cell areas, cell densities (cells per km^2), and the number of interfering tiers used in the simulations.

Table 2. ISD,electrical tilt, cell area, cell density and interfering tiers.

$\overline{d}_{site}$	ϕ_{etilt}	A_{cell} [km^2]	ρ_{cell} per km^2	Interfering tiers
960 m	3.5°	0.26	3.8	2
828 m	4.1°	0.2	5.1	2
593 m	5.8°	0.1	9.9	3
297 m	11.4°	0.03	39.3	4
170 m	47.5°	0.008	119.9	4

3. CAPACITY EFFICIENCY ANALYSIS

3.1. SINR evaluation and mapping to Shannon capacity

The performance of any cellular system or layout in a certain environment is highly dependent upon the radio propagation conditions. The quality of the radio link is determined by the coverage and the interference conditions which set a cap on the maximum throughput/users per cell, as defined by Shannon capacity bound, C:

$$C = W \log_2 (1 + \Gamma) \qquad (5)$$

where W is the bandwidth of the system, Γ is the signal-to-interference-noise ratio (SINR), which defines the radio propagation condition. From (5) it is evident that the cell/area spectral efficiency depends directly on Γ.

Assuming that there is no intra cell interference (typical assumption for frequency reuse 1 systems like orthogonal frequency division multiple access (OFDMA), where perfect orthogonality is assumed between the users of the same cell), the SINR at a jth receiver point (both outdoor and indoor) is calculated using the following relation:

$$\Gamma_j = \frac{S_{j,own}}{\sum I_{j,other} + P_n} \qquad (6)$$

where $S_{j,own}$ is the received signal power from the own cell (serving cell) at j^{th} receiver point, $I_{j,other}$ is the received interference power from the other cells at the jth receiver point, and P_n is the noise floor level which includes the noise figure of the receiver as well.

In a multi-cellular scenario, a cell having the strongest signal level is considered as the serving cell and the rest are treated as interferers. For a set of i cells reachable at the jth receiver, the best serving signal can be found as:

$$S_j = \arg\max_i \left(\Pr_{0j}, \Pr_{1j}, ..., \Pr_{ij} \right) \tag{7}$$

where, Pr_{ij} is the received signal power from the ith center cell site at j^{th} receiver.

3.2. Cell spectral efficiency and area spectral efficiency

We consider a fully loaded scenario in our simulation, i.e. all the base stations are transmitting at full power at all times. The *cell spectral efficiency,* η_{cell}, is defined as the maximum bit rate per Hertz that a cell can support under certain radio propagation conditions. For an area with a cell density of ρ_{cell} and average cell spectral efficiency of $\overline{\eta}_{cell}$, the *average area spectral efficiency,* $\overline{\eta}_{area}$, for a fully loaded system with constant interferers is defined as [28]:

$$\overline{\eta}_{area} \left[bps / Hz \ per \ km^2 \right] = \rho_{cell} \times \overline{\eta}_{cell} \tag{8}$$

where $\overline{\eta}_{cell} = \left\langle \dfrac{C}{W} \right\rangle$.

3.3. Capacity efficiency result and analysis

The general target of radio network planning is to design a network that provides sufficient coverage and maximizes overall capacity of the network with minimal costs. One of the most obvious methods for enhancing the network capacity is to increase the number of cells. However, the achievable average SINR ratios that eventually define the capacities (or 'cell spectral efficiency') in the cell level depend heavily on the network configuration. In this chapter, the results from the simulations, in terms of coverage, radio channel conditions and capacity, are analyzed and discussed.

Figure 2 provides the radio coverage statistics for the center cell site for both a) outdoor and b) indoor receiver points, for different ISDs. For the outdoor environment we can see that the signal levels improve in the whole cell area as macrocellular network is densified. Further densifying the network to ISD 170 m is shown to slightly degrade the coverage due to heavy downtilt angle (47.5°). For the indoor environment, the coverage pattern follows the same as in the outdoor. Again in the extreme densification scenario (ISD 170 m), the coverage performance degrades even beyond ISD 969 m.

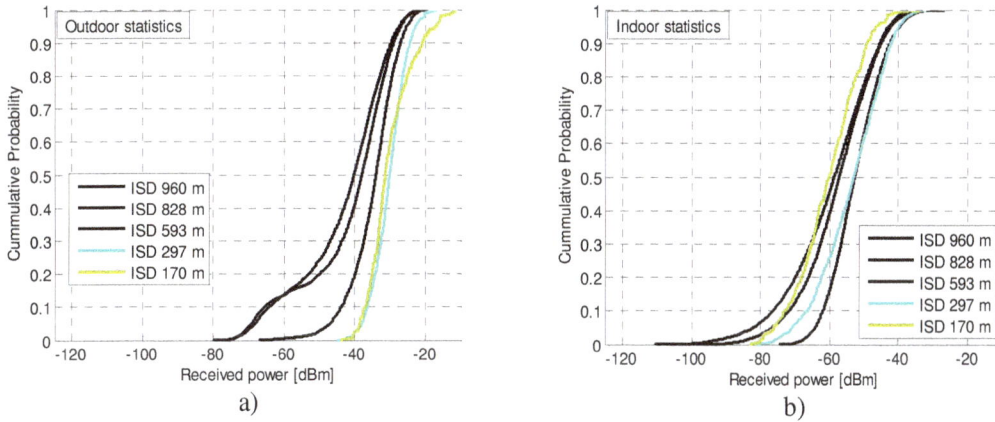

Figure 2. CDF distribution of received signal strength for a) outdoor receiver pointsa and b) indoor receiver points.

Figure 3 gives the center cell site statistics for the radio channel conditions (SINR) for both the outdoor (a) and indoor environment (b). In the outdoor environment, the radio channel conditions does not improve much in the overall cell, rather the SINR performance starts to degrade as we densify the network. The reduction in the SINR performance is more visible in the indoor environment with ISD 170 m having the worst interference conditions of all due to close proximity of the interfering cells.

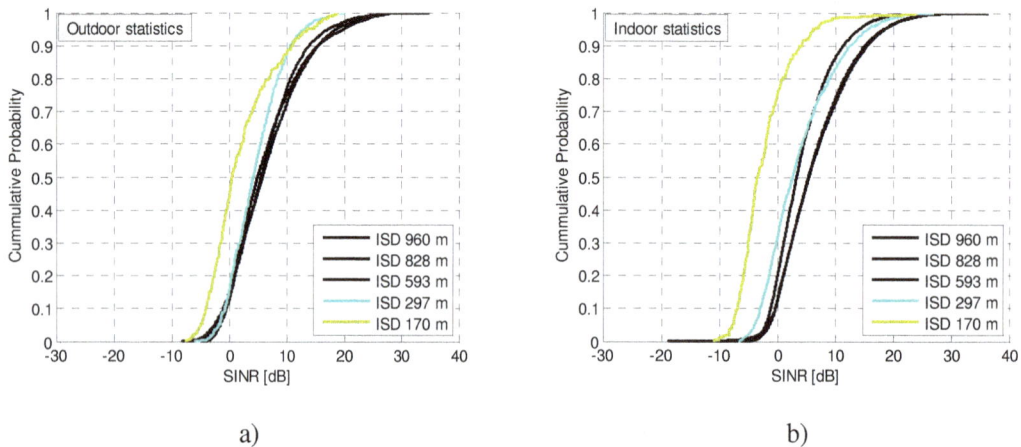

a)   b)

Figure 3. CDF distribution for SINR [dB] for a) outdoor receiver points and b) indoor receiver points.

From the overall cell site capacity perspective, the improvement in the cell edge performance is of significance, as these regions, due to being away from the serving base station, experience worse radio propagation conditions. Hence, it is extremely important that the cell edge conditions in terms of cell capacities are as high as possible. The maximization of these edge conditions can be achieved with a proper radio network deployment that eventually minimizes the inter-cell interference caused by the overlap between adjacent or neighboring cells. Thus, in our analysis we will focus on the lower 10th percentile values, which represent the conditions at the cell edge regions.

Figure 4 shows the statistical 10th percentile values for the received signal levels (i.e., coverage) and SINR, respectively, for the outdoor (a) and different indoor floor levels (b). The x-axis indicates the cell density per km^2 and y-axis the corresponding received signal strength [dBm] or SINR [dB]. For analysis, we have grouped the indoor floor levels into three classes; the bottom floors, middle floors and the top floors. The bottom floors bar presents the average of the 10th percentile values on the 1st and the 2nd floor, the middle floors bar indicate the average of the 10th percentile values on the 4th and the 5th floor, while the top floors bar shows the average of the 10th percentile values on 7th and 8th floor. From Figure 4a) we can see that the outdoor receiver points experience quite high signal levels from the very beginning as compared to the indoor floors. The received signal levels are relative to receiver noise floor level which is at -92 dBm (as shown by the dashed line). For less densified configurations, the receiver points in the lower floors experience high signal losses as compared to ones on the top floors. However, as a result of densification of the network, the overall coverage levels start to improve. The improvement in the coverage level comes from the deployment of more base stations together with antenna down tilt that results in smaller cell sizes, thereby reducing the path losses. Subsequent densification of the network does not bring any further improvement in the indoor coverage, while the outdoor receiver points experience a moderate improvement in the average signal levels. In the extreme case of 120 cells/km^2 (or average ISD of 170 m), the average signal levels saturate for receivers in outdoor and top floors, whilst the signal levels in the middle and lower floors start to experience coverage limitations. This is due to very high antenna tilt angles that cause signal losses in the lower floors.

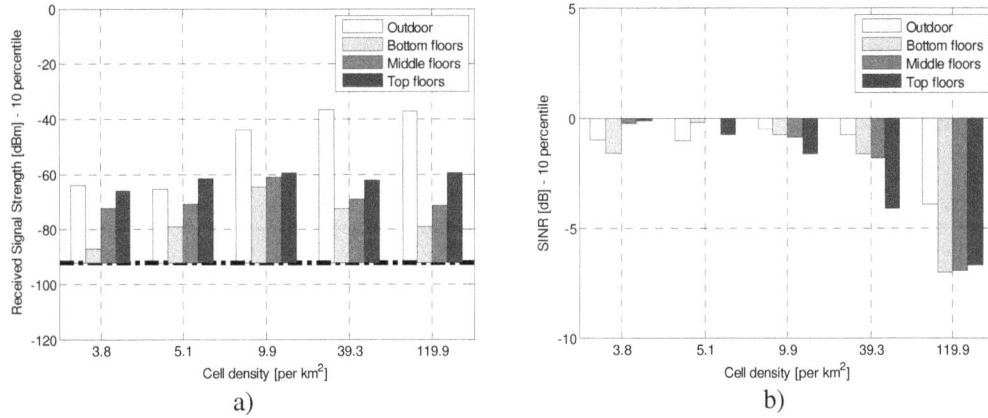

a)   b)

Figure 4. Cell edge values (the 10th percentile statistics) for a) received signal strength [dBm] and b) SINR [dB].

Figure 4b) presents the 10th percentile values i.e., the statistics from the cell edge, for SINR in outdoor and indoor environment for different cell densities. Although the coverage conditions on the top floor are better than in the middle and lower floors, the SINR performance degrades quite abruptly on top floor as the cell density increases. This is due to the rising interference conditions that become more prominent on the top floors as the network is densified. On the other hand, as a result of coverage improvement, the radio conditions in the lower and middle floor improve slightly when the network is densified to the level of 5 cells/km^2 (or average ISD of 828 m). For more densified configurations, lower and middle floors start to become coverage and clearly interference limited.

Table 2. Capacity performance at the cell border region (the 10th percentile values) for outdoor and different indoor floor levels.

		Cell spectral efficiency, η_{cell}, [bps/ Hz]			
		10 percentile			
$\bar{d}_{site}$	ρ_{cell} per km^2	Outdoor	Bottom floors	Middle floors	Top floors
969 m	3.8	0.84	0.76	0.96	0.98
828 m	5.1	0.84	0.97	1	0.88
593 m	9.9	0.92	0.88	0.86	0.75
297 m	39.3	0.88	0.75	0.73	0.47
170 m	119.9	0.49	0.26	0.27	0.28

Table 4 provides the 10th percentile values for the cell spectral efficiency versus cell densities, for the outdoor and different indoor floor levels. The SINR values under the dominance area of the center site are directly mapped to the cell spectral efficiency. In a full load condition, the cell efficiency is shown to decrease as the network is densified. Initially (3.8 cells/km^2), the cell edge spectral efficiency is at the level of 0.84 bps/Hz and reduces to the level of 0.49 bps/Hz for outdoor locations when network is densified to the level of 120 cells/km^2. For the indoor floor levels, the overall cell edge efficiency is higher on the middle and top floors as compared to the lower floor levels and even outdoor location. However, as the network is densified to 120 cells/km^2, the cell spectral efficiency reduces to approximately 0.27 bps/Hz on all the floor levels.

The higher degree of resource reuse due to denser deployments results in an increase of the area spectral efficiency as shown in Table 5. The impact of outdoor and indoor location on the area spectral efficiency is observed to be quite marginal in the beginning (ISD of 969m and 828m), but as the network is densified, the difference in the area capacity gain starts to become more visible. For the cell spectral efficiency, the effect tends to get more recognizable when the network is densified beyond the level of 5 cells/km^2 (or average ISD of 828 m). This is attributed to the rising indoor interference level, mostly on the top floors as shown in Figure 2b).

Table 3. Average cell and area spectral efficiency for different ISDs.

$\overline{d}_{site}$	ρ_{cell} $per\ km^2$	$\overline{\eta}_{cell}[bps\ /\ Hz]$		$\overline{\eta}_{area}[bps\ /\ Hz\ /\ km^2]$	
		Outdoor	Indoor	Outdoor	Indoor
969 m	3.8	2.7	2.67	15.1	14.96
828 m	5.1	2.65	2.61	22.42	22.06
593 m	9.9	2.57	2.05	36.05	28.74
297 m	39.3	2.09	1.99	92.81	88.06
170 m	119.9	1.65	0.88	289.2	153.9

In mobile communications industry, it has been widely speculated that more than 70% of the overall network traffic originates from indoor users. Hence, to properly dimension its network a mobile operator has to consider service provisioning from the indoor perspective. However, the results indicate that macrocellular network densification in urban Manhattan environment clearly suffers from inefficiency indoors. If the radio network planning target is limited to coverage provisioning for outdoor users only, the densification efficiency is higher (see Figure 3a). On the other hand, if networks are planned for indoor coverage (as in practice), the efficiency is clearly lower. To illustrate this for a practical outdoor/indoor user distribution, we consider a scenario where majority of the receiver points are located indoors i.e., we assume the receiver distribution ratio as 20% outdoor and 80 % indoor. To ensure statistically reliable results, the receiver points were randomly selected, with several iterations, from both outdoor and indoor environment with the intended ratio.

Figure 5 illustrates the capacity analysis in a slightly different way, where the relative area spectral efficiency for a network with different cell densities per km^2 has been depicted. The area capacity values are relative with respect to nominal site density (3.8 cells/km^2). The dashed line illustrates 100% densification efficiency (ρ_{eff}) line, whereas the solid line shows the improvement of the area spectral efficiency for 20/80 % outdoor/indoor receiver point distribution. For less dense configuration, there can be observed a linearly increasing trend in the area spectral efficiency. The densification efficiency is still roughly 0.8 for 9.9 cells/km^2 (or average ISD of 597 m). However, beyond that point the efficiency can be observed to deteriorate significantly due to increase of inter-cell interference resulting in from network densification, and abruptly drops down to 0.38 for 119.9 cells/km^2 scenario. These results clearly illustrate the inefficiency of macrocellular network densification with a more practical user distribution.

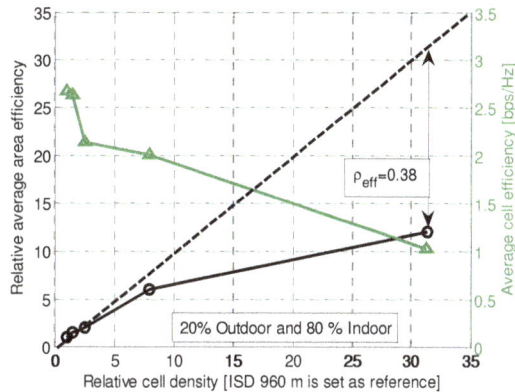

Figure 5. Relative area efficiency (○) and average cell efficiency (△) vs. relative cell density (the dashed line indicate a linearly increasing area efficiency curve in an ideal case).

4. ENERGY EFFICIENCY ANALYSIS

4.1. Power consumption modeling

In order to estimate the energy efficiency of a macrocellular network over a certain area, it is important that the power consumption of a single base station (BS) site is modelled as accurately as possible. A base station site comprises of a base station unit, also known as the base transceiver station (BTS), which has the capability to transmit and receive radio signals to and from the mobile subscribers. It acts an interface between a subscriber and the core network, enabling the subscriber to wirelessly connect to the mobile operator's network. A macrocellular base station consists of various internal power consuming components that contribute to the overall energy usage. The components are power amplifier, transceiver, digital signal processing (DSP) chips and rectifier [14], [21]. Each sector of a base station has its own set of these components, hence the total power consumption scale with the number of sectors per base station. Apart from the internal components, there are some external equipment that also have significant contribution to the overall power usage of a macro base station cell site. The base stations require these external equipment for its smooth operation and connectivity to the mobile operator's core network, for instance, an air conditioning unit is required to maintain an ambient temperature (usually 25°C) inside the base station shelter which houses the macro base station unit. Moreover, for connecting the base station to the backhaul network a transmission unit is installed. Note that the choice of transmission unit has considerable impact not only on the power consumption but also on the capital expenditure CAPEX (cost of equipment, deployment of backhaul network to the base station site etc.) and operating cost OPEX (power consumption, leased line rent etc.). For example, an optical transmission unit will consume less power and provides a very high capacity than a microwave transmission unit, but the associated cost of leasing or deploying an optical network increases the CAPEX considerably. In our study we consider a microwave link transmission unit, which is still widely used solution for mobile backhaul connectivity. For accurate estimation of the total network energy utilization, it is essential to also take into account the contributions from these external equipment.

An analytical power consumption model for a macrocellular base station site is proposed in [13] that takes into account the contribution from both internal base station components and external site equipment. The model further takes into account the impact of hourly network load on the total base station power utilization as well. As such the power consumption of a macro base station site, $P_{consumption/Macro}$, is given by [21]:

$$P_{consumption/Macro}\left[Watt\ hour\right] = P_{constant} + P_{load} \cdot F \qquad (9)$$

where $P_{constant}$ is the contribution from the internal components and external equipment whose power consumption is network load independent, P_{load} is the contribution from the components/equipment whose power consumption depends on the instantaneous load of the network, and F is the load factor varying from 0 to 1, with 0 meaning 'no load' and 1 pertains to 'high load/full load'. The F scales the power consumption of the load dependent base station components according to the network load per hour. In our analysis we use $F = 1$, which is the worst case scenario, i.e. the network is running on a 100 % load and all the load dependent components are consuming high power.

The load independent power consumption components include; rectifier (*Rect*), microwave link (*MLink*) and air-conditioning (*Air-cond*) unit. Thus, $P_{contant}$ is given by:

$$P_{constant}[Watts] = \left(n_{sector} \cdot P_{Rect}\right) + P_{MLink} + P_{Air-Cond} \qquad (10)$$

where P_{Rect} is the power consumption of the rectifier, P_{MLink} is the power consumption of a microwave link, and $P_{Air-cond}$ is the contribution from the air conditioning unit. As every sector of a macro base station has a separate rectifier, the total power consumption of the rectifier is scaled with the number of sectors, n_{sector}, installed at a base station.

The network load dependent power consumption components include power amplifier, transceiver and a baseband processing card (digital signal processing, DSP card). Thus, P_{load} is given by:

$$P_{load}[Watts] = n_{sector} \cdot \left[n_{Ant}\left(P_{Amp} + P_{TRX}\right) + P_{DSP} \right] \qquad (11)$$

where P_{Amp} is the power consumption of the amplifier, P_{TRX} is the power consumption of a transceiver and P_{DSP} is the power usage contribution from the DSP card. The power contribution from these components is scaled with n_{sector}. Note that in case of MIMO (multiple input multiple output) configuration, a sector can have more than one antenna installed. Each antenna has a separate amplifier and a transceiver. Hence, the number of antennas per sector, n_{Ant}, also has an effect on the total power consumption. In our analysis we assume a SISO (single input and single output) configuration, i.e. 1 antenna per sector. The power consumption of the amplifier, P_{Amp}, depends on the input power requirement of the antenna, P_{TX}, and efficiency of the power amplifier, η_{amp} and is given by:

$$P_{Amp} = \frac{P_{TX}}{\eta_{amp}} \qquad (12)$$

Table 6 summarizes the input parameters for the base station power consumption model. The parameters are approximate values of the power consumption of different base station components and external equipment taken from [21], except for the power amplifier efficiency value which is typically 30 % to 55% for the base stations. Using the input values in Table 3, the power consumption of a 3-sectored macro cellular base station operating in a full load condition is approximated to be 1338 Watts.

Table 6. Input parameters for the power consumption model.

Component/Equipment	Unit	Value
Number of sectors, n_{sector}		3
Number of antennas per sector, n_{Ant}		1
Transmit power at the antennas, P_{TX}	[Watts]	20
Power consumption of DSP chip, P_{DSP}	[Watts]	100
Power Amplifier efficiency, η_{Amp}	[%]	45
Power consumption of Transceiver, P_{TRX}	[Watts]	100
Power consumption of Rectifier, P_{Rect}	[Watts]	100
Power consumption of Air-conditioning unit, $P_{Air\text{-}cond}$	[Watts]	225
Power consumption of Microwave-Link unit, P_{MLink}	[Watts]	80

4.2. Area power consumption and energy-efficiency

One of the most commonly used metric for assessing the energy efficiency of a network is by evaluating the energy-to-bit ratio performance of the network, i.e. the amount of energy consumed in transmitting one bit of information. On a network level, this relates to the total power consumed by the network in providing an aggregate network capacity. This methodology is appropriate for assessing the energy efficiency of a network operating under full load condition [13].

For studying the impact of base station site densification on the energy efficiency of the network, we first find the area power consumption by normalizing the total power consumption of a base station given in (9) to 1 km^2 area. The normalized area power consumption of a macro cellular base station, $P_{consumption/km^2}$, is given by:

$$P_{consumption/km^2}\left[Watts \ / \ km^2\right] = \frac{P_{consumption/Macro}}{A_{site}} \qquad (13)$$

where A_{Site} is the area of a base station (in km^2) and is defined as $3 \times A_{cell}$ for a 3 sectored base station. The area power consumption is deemed as an appropriate metric in a case where the network is operating below its full load capacity, and the target is to minimize the power consumption over an area [13]. Hence, we use the energy-bit-ratio metric in our analysis. Finally, we define the energy-efficiency as the power consumed in transmitting one bps/Hz and is calculated as following:

$$E_{eff}\left[bps \ / \ Hz \ / \ kW\right] = \frac{\overline{\eta}_{area}}{P_{consumption/km^2}} \qquad (14)$$

where $\overline{\eta}_{area}$ is the average area spectral efficiency as defined in (8).

4.3. Energy efficiency results and analysis

Table 7 summarizes the energy efficiency results for different inter site distances (varying cell densities) for outdoor and indoor receiver points. As we can note from the results, the power consumption per km^2 increases with the increase in the cell density. This is because the area power consumption depends on the coverage area of the base station. We assume that the network coverage is continuous, without any coverage holes, and that the base stations consume same amount of power irrespective of the coverage area size. Hence, densification of the network leads to increased power consumption per area proportionally with the increase in number of base stations.

Table 7. Area power consumption and energy efficiency for different ISDs.

$\overline{d}_{site}$	ρ_{cell} $per\,km^2$	$P_{consumption/km^2}[kW\,/\,/km^2]$	$\overline{\eta}_{area}[bps\,/\,Hz\,/\,km^2]$		$E_{eff}[bps\,/\,Hz\,/\,kW]$	
			Outdoor	Indoor	Outdoor	Indoor
969 m	3.8	1.7	15.1	14.96	8.9	8.8
828 m	5.1	2.3	22.42	22.06	9.7	9.6
593 m	9.9	4.4	36.05	28.74	8.2	6.5
297 m	39.3	17.6	92.81	88.06	5.3	5
170 m	119.9	53.5	289.2	153.9	5.4	2.9

By densifying the network, the spectrum resource are reused more frequently, which thereby improves the area spectral efficiency. However, looking at the impact of site densification on the energy efficiency of the network, it is noted that although increasing the number of bps/Hz/km^2, the energy needed to transmit 1 bps/Hz also increases as we densify our network, especially in the indoor environment. As an example, consider the initial case of 3.8 cells/km^2 (ISD 960 m), where the average area spectral efficiency is the same for both outdoor and indoor environment. In this case, the total power consumed per km^2 is approximately 1.7 kW, which leads to energy efficiency of approximately 8.9 bps/Hz/kW for outdoor and 8.8 bps/Hz/kW for indoor environment. Upon decreasing the inter-site distance to 828 m (i.e., 5.1cells/km^2), a slight improvement can be observed in the energy efficiency (9.7 bps/Hz/kW for outdoor and 9.6 bps/Hz/kW for indoor). This improvement comes from the fact that in the initial stages of densification, the macrocellular network is slightly coverage limited. Hence, by densifying the network, the coverage levels improve in both outdoor and indoor environment, thereby improving the radio channel conditions and hence permitting higher cell spectral efficiency. Subsequent densification down to ISD 593 m and 297 m starts to degrade the energy efficiency performance as the network becomes more and more interference limited. The impact of degradation is more visible in the indoor environment due to relatively low rate of spectral efficiency improvement as compared to the outdoor environment. Eventually, when we densify the network to an extreme case (ISD 170 m case or 120 cells/km^2), given approximately 32 times more cells/km^2 as compared to initial ISD 969 m case, the area power consumption increases significantly. However at this stage, a slight improvement in the outdoor energy efficiency can be observed, but for the indoor environment the degradation in the energy efficiency performance extends even further. The reason is attributed to the capacity inefficiency in macrocellular networks.

As discussed in the previous section, the macrocellular network densification suffers from capacity in-efficiency in the indoor environment. The relative indoor capacity gain that we can achieve when we densify the network from 3.8 cell/km^2 to approximately 120 cells/km^2 (32 times more cells) is only 38%, which means that although the power consumption per km^2 is increasing as we are increasing the number of cell density, the associated area capacity in the indoor environment to offset this increase of power consumption is not enough, hence the Watts/bit increases. Looking at the big picture, we conclude that pure macrocellular network densification suffer from both outdoor and indoor energy inefficiency.

5. COST EFFICIENCY ANALYSIS

One of the most obvious ways to increase the capacity of a wireless network is by reusing the existing allocated spectrum as frequently as possible throughout the network service area, in other words by increasing the base station density. As such, the capacity of a cellular network is considered to be proportional to the base station density. Unfortunately, the network infrastructure cost also increases with the number of base stations which is a key concern for cost aware mobile operators that are striving to provide better services at lower cost in a highly competitive market. For any technological pathway to be feasible, the benefits must outweigh the incurred costs. Hence, correct estimation of the benefits and costs is very crucial. Cost efficiency analysis, or cost-benefit analysis, is one of the key methodologies that provide a general picture of the cost structure of an evolutionary pathway for a certain technology or system and whether or not it is a feasible option for investment. In this section we describe the cost modelling methodology used in our analysis, and based on the cost model we evaluate the cost-bit ratio for different ISDs (different site densities). Finally, we conclude the section by analyzing and discussing the cost-efficiency results.

5.1. Cost modeling

The cost of deploying a macrocellular network can be broadly divided into two types:

   i.   Investment cost or CAPEX (capital expenditure),
   ii.   Running/operational costs or OPEX (operational expenditure)

The CAPEX consists of equipment costs like *radio base station, transmission equipment, antennas, cables*, and site build out and installation cost. OPEX consists of *site rental, electricity, transmission or leased line*, and *OA&M (operation, administration & maintenance)*. In addition to these, there can be cost components as such as radio network planning, core network and marketing costs whose impact can be modeled and taken into account as part of the radio network costs [29]. However, in the frame of this article, the scope is limited to items listed for CAPEX and OPEX as they typically depend very strongly on the number of deployed radio components. Combining CAPEX and OPEX gives the total cost of ownership (TCO) value of the deployed network.

The total cost structure of a mobile operator is dominated by the accumulated running costs i.e. the OPEX [30], which spans over the life-time of the network, while the CAPEX is considered during the initial network roll-out phase or when the network is upgraded. Thus, in order to account for both the CAPEX and OPEX in finding the 'total cost per base station' we use a standard economical method known as discounted cash flow (DCF) analysis, which gives the net present value (NPV) of the base station cost.

The net present value of the base station is simply found by summing up the discounted annual cash flow expenditure for a given study period (in years) [30, 31]. Mathematically;

$$BS_{NPV} = \sum_{i=0}^{Y-1} \frac{c_i}{\left(1+r\right)^i} \tag{15}$$

where Y is the study period in years (typically 8 years for base stations value depreciation), c_i is the total annual expenditure per base station (total annual cost which includes running cost and may include investment cost) in the ith year and r is the discount rate which is assumed to be equal to 10%. Table 8 gives the various cost items related to CAPEX and OPEX and their approximate values. The values have been adopted from [31-33].

To find the net present value of the base station we make the assumption that the mobile operator is deploying its network as a Greenfield project[1] and that the whole network is deployed in the first year, so the CAPEX will be considered only for first year. Using the values in Table 8, the total cost per base station is 93 k€.

In order to assess the cost viability of the macrocellular network densification as an evolutionary pathway for future mobile broadband systems, we analyze the cost-bit ratio efficiency, or simply the cost-efficiency metric for the different ISDs. The cost efficiency is defined as the cost incurred in transmitting one bit/Hz and is calculated as following:

$$c_{eff}\left[bps\,/\,Hz\,/\,k€\right] = \frac{\overline{\eta}_{area}}{T_{cost/km^2}} \qquad (15)$$

where T_{cost/km^2} is the total area cost, i.e. the total cost of base stations over 1 km^2 area and $\overline{\eta}_{area}$ is the average area spectral efficiency.

Table 8. CAPEX and OPEX related cost for a macro cellular base station.

CAPEX (Initial costs)	
Macro base station equipment	10 k€
Site deployment cost	5 k€
Total CAPEX	15 k€
OPEX (Running costs)	
Site rent (lease)	5 k€/year
Electricity (power consumption charges)[2]	2.25 k€/year
Transmission line / Leased line rent[3]	0 k€/year
Operation, Administration & Maintenance (OA&M)	5 k€/year
Total OPEX per annum	12.25 k€

5.2. Cost efficiency results and analysis

The cost efficiency analysis results for different macrocellular cell densities have been summarized in Table 9. As evident, the total cost of deployment per km^2 increases as we increase the base station density. However, the important metric to investigate is not the aggregate cost but the cost per bit efficiency, i.e., we are interested in the relative gain that we can achieve from densification. In other words, we need to investigate whether the macrocellular densification can provide decent capacity gain to offset the incurred cost of deployment and hence bring down the cost per bit to make macrocellular densification a viable business case for investment.

In general, looking at the cost efficiency values in Table 9, it can be seen that it follows the energy efficiency performance pattern. In the initial stages (increasing the cell density from 3.8 to 5.1 cells/km^2), there is a slight improvement in the cost efficiency performance for both outdoor and indoor environment. However, further densification not only degrades the cost efficiency performance but also the difference between outdoor and indoor environment starts to become more noticeable. This is, as mentioned previously, attributed to the inefficiency of pure macrocellular network densification in the indoor environment. From (15) it can be seen that the

---

[1] Any new network which is designed and deployed from scratch, i.e. operator has no prior deployment in that region.

[2] The annual cost of electricity consumption is normally calculated based on total kilowatt hours (kWh) consumed during a given year. The cost per electricity unit is assumed to be 0.2 €/kWh. The calculated annual cost of electricity for macrocell base station given in Table 5 conforms quite well to the range given in [32].

[3] We assume that the base stations are connected to the backhaul network via a microwave transmission link. Thus, there is no leased line (E1/T1) rent or deployment cost of fiber optic network.

cost efficiency depends on the area spectral efficiency. The more the network is densified, the smaller capacity gain in the indoor environment is achieved, which results in higher cost per bit. For the outdoor environment, increasing the base station density to the extreme case (i.e., 120 cells/km^2) results in relatively higher area capacity gain than the indoor environment, which results in slightly improved cost efficiency. However, the efficiency still lags behind the cost efficiency of ISD 969 m. Hence, it can be concluded that the pure macrocellular densification suffers from cost inefficiency especially in indoor environment in dense urban area.

Table 9. Total area cost and cost efficiency results for different ISDs.

$\bar{d}_{site}$	ρ_{cell} per km^2	$T_{cost/km^2}\left[k\epsilon\,/\,km^2\right]$	$\bar{\eta}_{area}[bps\,/\,Hz\,/\,km^2]$		$c_{eff}[bps\,/\,Hz\,/\,k\epsilon]$	
			Outdoor	*Indoor*	*Outdoor*	*Indoor*
969 m	3.8	117	15.1	14.96	0.13	0.13
828 m	5.1	157	22.42	22.06	0.14	0.14
593 m	9.9	305	36.05	28.74	0.12	0.09
297 m	39.3	1224	92.81	88.06	0.08	0.07
170 m	119.9	3720	289.2	153.9	0.08	0.04

6. CONCLUSION

In this paper we have shown how macrocellular network densification reduces the cell spectral efficiency under full load conditions and varying receiver point distributions (outdoor/indoor) in dense urban environment using a 3D radio signal propagation model. As a result of reduction of cell spectral efficiency, the area spectral efficiency starts to saturate and macrocellular network densification becomes less efficient. However, if an operator targets only for outdoor coverage with macrocellular network, it still might be sufficient to fulfill the capacity demands through densification. From indoor coverage and capacity provisioning point of view, however, macrocellular network densification is clearly less efficient (0.38 densification efficiency with 20/80% outdoor/indoor receiver point distribution). The reduction in the spectrum efficiency also has a direct impact on the energy and cost efficiency of macrocellular network: lower spectrum efficiency in the indoor environment results in higher energy consumption and cost per bit. Moreover, from coverage point of view, in a dense urban environment with high rise buildings, the macrocellular network is not efficient in providing good coverage to indoor floors. This affects the attainable capacity in the indoor environment, and clearly further indicates that increasing network capacity demand will require alternative deployment strategies, as introduction of small cells (micro) or indoor (pico, femto), that will provide local indoor capacity within the network. On the macrocellular network, however, alternative mechanisms as interference mitigation techniques using smart antenna systems, base station transmission coordination or interference cancellation mechanisms are clearly needed to increase the densification efficiency.

Future work will concentrate on analyzing the coverage, capacity, costs and energy consumption of small or microcell networks and comparing that with macrocellular networks. Moreover, we will evaluate the effect of macrocellular network densification with base station coordination from capacity, costs, and energy-efficiency point of views.

ACKNOWLEDGEMENTS

This research work has been financially supported by the Finnish Agency for Technology and Innovation (Tekes), under the Sino-Finland collaboration project "Energy-Efficient Wireless Networks and Connectivity Devices – Systems (EWINE-S)".

REFERENCES

[1]   *Traffic and Market Data Report*, (2012), Annual report, Ericsson Inc.

[2]   Macdonald, V. H., (1979), "Advanced Mobile Phone Service: The Cellular Concept", *The Bell System Technical Journal*, Vol. 58.

[3]   Liang, Y., et al., (2008), "Evolution of Base Stations in Cellular Networks- Denser Deployment versus Coordination", IEEE *International Conference on Communications*.

[4]   Badic, B., Farrell T. O', Loskot, P., He J., (2009), "Energy Efficient Radio Access Architectures for Green Radio: Large versus Small Cell Size Deployment", IEEE *70th Vehicular Technology Conference (VTC)*.

[5]   Richter, F., Fettweis, G., (2010), "Cellular Mobile Network Densification Utilizing Micro Base Stations", IEEE *International Conference on Communications*.

[6]   Hiltunen, K., (2011), "Comparison of Different Network Densification Alternatives from the LTE Uplink Performance Point of View", IEEE *74th Vehicular Technology Conference (VTC)*.

[7]   Hiltunen, K., (2011), "Comparison of Different Network Densification Alternatives from the LTE Downlink Performance Point of View", IEEE *International Symposium on Personal, Indoor and Mobile Radio Communications (PIMRC)*.

[8]   *ICT Sustainability Outlook: An Assessment of the Current State of Affairs and a Path Towards Improved Sustainability for Public Policies*, White paper, (2013), BIO Intelligence services and Alcatel Lucent.

[9]   *Sustainable Energy Use in Mobile Communications*, (2007), White paper, Ericsson Inc.

[10]   Pickavet M., et al., (2008), "Worldwide Energy Needs for ICT: The Rise of Power-Aware Networking", *International symposium on Advanced Networks and Telecommunications Systems*.

[11]   Richter F., et al., (2009), "Energy Efficiency Aspects of Base Station Deployment Strategies for Cellular Networks", IEEE *70th Vehicular Technology Conference (VTC)*.

[12]   EU Commissioner calls on the ICT industry to reduce its carbon footprint by 20% as early as 2015, MEMO/09/140. Press release, *EU Commission* (www.europa.eu)

[13]   Correia, L.M., et al., (2010), "Challenges and Enabling Technologies for Energy Aware Mobile Radio Networks", IEEE *Communications Magazine*.

[14]   Auer G., et al., "Cellular Energy Efficiency Evaluation Framework", IEEE *73rd Vehicular Technology Conference (VTC)*.

[15]   Chen T., et al., (2010), "Energy Efficiency Metrics for Green Wireless Communications", IEEE *International conference on Wireless Communications and Signal Processing (WCSP)*.

[16]   Ericson M., (2011), "Total Network Base Station Energy Cost vs. Deployment", IEEE *73rd Vehicular Technology Conference (VTC)*.

[17]   Le T., and Nakhai M., (2010), "Possible power-saving gains by dividing a cell into tiers of smaller cells", IET *Electroncis Letters*, Vol. 46, No. 16.

[18]   Leem H., Baek S. Y. and Sung D. K., (2010), "The Effects of Cell Size on Energy Saving, System Capacity, and Per-Energy Capacity", IEEE *Wireless Communications and Networking Conference*.

[19]   Tombaz S., Sung K. W. and Zander J., (2012), "Impact of Densification on Energy Efficiency in Wireless Access Networks", IEEE *Globecom Workshops.*

[20]   Tombaz S., et al., (2011), "Impact of backhauling power consumption on the deployment of heterogeneous mobile networks", IEEE Globecom conference.

[21]   Deruyck M., Joseph W., and Martens L., (2012), "Power consumption model for macrocell and microcell base stations", *Transactions on Emerging Telecommunications Technology.*

[22]   Schuster, J. and Luebbers, R., (1996), "Shooting and Bouncing Rays: Calculating the RCS of an arbitrarily shaped cavity", IEEE *Transactions on Antenna and Propagation*, Vol. 1.

[23]   Keller, J. B., (1996), "Geometrical Theory of Diffraction", *Journal of Optical Society of America*, Vol. 52, No. 2.

[24]   Kouyoumjian, R. G., Pathak, P. H., (1974), "A Uniform Geometrical Theory of Diffraction for an edge in a perfectly conducting surface", IEEE *Proceedings*, Vol. 62, No. 11.

[25]   Balanis, C., (1989), *Advanced Engineering Electromagnetics*, John Wiley & Sons Inc.

[26]   Gunnarsson, F., et al., (2008), "Downtilted Base Station Antennas - A Simulation Model Proposal and Impact on HSPA and LTE Performance", IEEE 68th *Vehicular Technology Conference (VTC).*

[27]   Sridhara V., and Bohacek S., (2007), "Realistic propagation simulation of urban mesh networks", International Journal of Computer and Telecommunications network Computer Networks and ISDN Systems (COMNET).

[28]   Alouni M., and Goldsmith A., (1999), "Area Spectral Efficiency of Cellular Mobile Radio Systems", IEEE *Transactions on Vehicular Technology*, Vol. 48, No. 4, pp. 1047-1066

[29]   Giles. T, et al., (2004), "Cost Drivers and Deployment Scenarios for Future Broadband Wireless Networks", IEEE 59th *Vehicular Technology Conference (VTC).*

[30]   Smura T., (2012), *Techno-economic modelling of wireless network and industry architectures,* Doctoral dissertation, Dept. Communications and Networking, Aalto University.

[31]   Johansson K., et al., (2004), "Relation between base station characteristics and Cost structure in Cellular systems". IEEE *International Symposium on Personal, Indoor and Mobile Radio Communications (PIMRC).*

[32]   Markendahl, J. and Ma☐kitalo, O., (2010), "A comparative study of deployment options, capacity and cost structure for macrocellular and femtocells networks", IEEE *International Symposium on Personal, Indoor and Mobile Radio Communications (PIMRC).*

[33]   Niemela J., and Isotalo T., (2012), *Coverage constrained Techno-economical comparison of Macro and Small Cell Planning Strategies,* Internal report, Dept. Electronics and Communications Engineering., Tampere University of Technology.

7

GROUP SESSION KEY EXCHANGE MULTILAYER PERCEPTRON BASED SIMULATED ANNEALING GUIDED AUTOMATA AND COMPARISON BASED METAMORPHOSED ENCRYPTION IN WIRELESS COMMUNICATION (GSMLPSA)

Arindam Sarkar[1] and J. K. Mandal[2]

[1]Department of Computer Science & Engineering, University of Kalyani, W.B, India
arindam.vb@gmail.com
[2]Department of Computer Science & Engineering, University of Kalyani, W.B, India
jkm.cse@gmail.com

ABSTRACT

In this paper, a group session Key Exchange multilayer Perceptron based Simulated Annealing guided Automata and Comparison based Metamorphosed encryption technique (GSMLPSA) has been proposed in wireless communication of data/information. Both sender and receiver uses identical multilayer perceptron and depending on the final output of the both side multilayer perceptron, weights vector of hidden layer get tuned in both ends. As a results both perceptrons generates identical weight vectors which is consider as an one time session key. In GSMLPSA technique plain text is encrypted using metamorphosed code table for producing level 1 encrypted text. Then comparison based technique is used to further encerypt the level 1 encrypted text and produce level 2 encrypted text. Simulated Annealing based keystream is xored with the leve2 encrypted text and form a level 3 encrypted text. Finally level 3 encrypted text is xored with the MLP based session key and get transmitted to the receiver. GSMLPSA technique uses two keys for encryption purpose. SA based key get further encrypted using Automata based technique and finally xored with MLP based session key and transmitted to the receiver. This technique ensures that if intruders intercept the key of the keystream then also values of the key not be known to the intruders because of the automata based encoding. Receiver will perform same operation in reverse order to get the plain text back. Two parties can swap over a common key using synchronization between their own multilayer perceptrons. But the problem crop up when group of N parties desire to swap over a key. Since in this case each communicating party has to synchronize with other for swapping over the key. So, if there are N parties then total number of synchronizations needed before swapping over the actual key is $O(N^2)$. GSMLPSA scheme offers a novel technique in which complete binary tree structure is follows for key swapping over. Using proposed algorithm a set of N parties can be able to share a common key with only $O(log_2 N)$ synchronization. Parametric tests have been done and results are compared with some existing classical techniques, which show comparable results for the proposed technique.

KEYWORDS

Metamorphosed, Automata, Session Key

1. INTRODUCTION

The sturdiness of the key is calculated in terms of linear complexity, randomness and correlation immunity. To devise a key following 4 basic features like large period of key, large linear complexity of the key, good random key sequence, high order of correlation immunity of the key sequence is required. The most important hazard for private key cryptography is how to firmly swap over the shared secrets between the parties. As a result, key exchange protocols are mandatory for transferring private keys in a protected manner. The first available key exchange protocol is known as Diffie-Hellman key exchange and it depends on the stiffness of computing discrete logarithms [1]. These days a range of techniques are available to preserve data and information from eavesdroppers [1]. Each algorithm has its own advantages and disadvantages. Security of the encrypted text exclusively depends on the key used for encryption. In cryptography the main security intimidation is man-in-the-middle attack at the time of exchange the secret session key over public channel.

In recent times wide ranges of techniques are available to protect data and information from eavesdroppers [1, 2, 3, 4, 5, 6, 7, 8, 9]. These algorithms have their virtue and shortcomings. For Example in DES, AES algorithms [1] the cipher block length is nonflexible. In NSKTE [4], NWSKE [5], AGKNE [6], ANNRPMS [7] and ANNRBLC [8] technique uses two neural network one for sender and another for receiver having one hidden layer for producing synchronized weight vector for key generation. Now attacker can get an idea about sender and receiver's neural machine because for each session architecture of neural machine is static. In NNSKECC algorithm [9] any intermediate blocks throughout its cycle taken as the encrypted block and this number of iterations acts as secret key. If n number of iterations are needed for cycle formation and if intermediate block is chosen as an encrypted block after $n/2^{th}$ iteration then exactly same number of iterations i.e. $n/2$ are needed for decode the block which makes easier the attackers life. As the same time as key exchange protocols are developed for exchanging key between two parties, many applications do necessitate the need of swapping over a secret key among group of parties. A lot of proposals have been proposed to accomplish this goal. As Multilayer perceptron synchronization proposal is a fresh addition to the field of cryptography, it does not provide a group key exchange mechanism. To solve these types of problems in this paper we have proposed a key swap over mechanism among cluster of multilayer perceptrons for encryption/decryption in wireless communication of data/information. This scheme implements the key swap over algorithm with the help of complete binary tree which make the algorithm scales logarithmically with the number of parties participating in the protocol. In this paper, additional sub key generation mechanism has been also proposed using local random search algorithm i.e. Simulated Annealing (SA) for encryption. This SA based key generation methods generates key which satisfies all 4 basic features. Proposed technique is also better than SATMLP [17] algorithm because of Metamorphosed guided Automata based comparison encryption rather than simple triangularized encryption in case of SATMLP [17].

The organization of this paper is as follows. Section 2 and 3 of the paper deals with character code table and metamorphosed code generation. Proposed encryption and decryption has been discussed in section 4 and 5. SA based key generation discussed in section 6. Section 7 deals with MLP based session key generation. GSMLPSA password based certificate generation scheme is given in section 8 and 9. Complexity analysis of the technique is given in section 10. Experimental results are described in section 11. Analysis of the results presented in section 12. Analysis regarding various aspects of the technique has been presented in section 13. Conclusions and future scope are drawn in section 14 and that of references at end.

2. CHARACTER CODE TABLE GENERATION

For plain text "tree" figure 1 shows corresponding tree representation of probability of occurrence of each character in the plain text. Characters 't' and 'r' occur once and character 'e' occurs twice. Each character code can be generated by travelling the tree using preorder traversal. Character values are extracted from the decimal representation of character code. Left branch is coded as '0' and that of right branch '1'. Table 1 shows the code and value of a particular character in the plain text.

Table1. Code table

Plain text	Code	Value
t	10	2
r	11	3
e	0	0

3. METAMORPHOSED CHARACTER CODE TABLE GENERATION

From the original tree metamorphosed tree is derived using mutation. Figure 2, 3 and 4 are the metamorphosed trees. After mutation new code values as obtained are tabulated in table 2. Tree having (n-1) intermediate nodes can generate 2^{n-1} mutated trees. In order to obtain unique value, the code length is added to the character if the value is identical in the table.

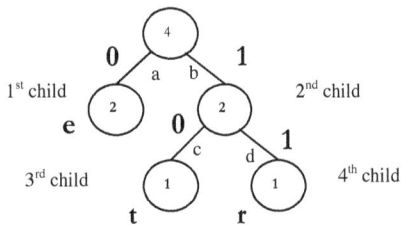

Figure 1. Character Code Tree

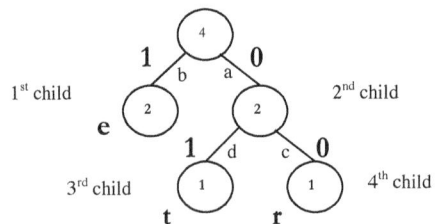

Figure 2. Swap the edges between a and b and between c and d.

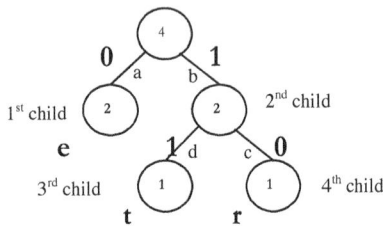

Figure 3. Swap the edges between c and d.
Edges a and b get unaltered.

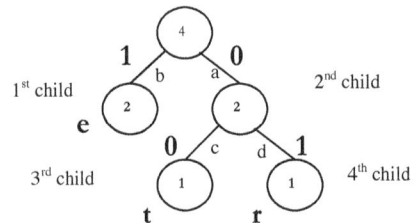

Figure 4. Swap the edges between a and b.
Edges c and d get unaltered.

Table2. Mutated code table

Character	Code	Value	Code	Value	Code	Value
t	01	1	11	3	00	0
r	00	0	10	2	01	1
e	1	2	0	0	1	2

4. ENCRYPTION ALGORITHM

Step 1. Plain texts are firstly encoded using the help of metamorphosed code table to get an level1 encoded text for enhancing the security perform the following comparisons based encryption for each block to get level2 encoded text.

> **Step 1.1** For the characters in the block perform following calculations. For the very first character in the block weight value of the previous character will be 0.
>
> $$Char\_Weight = Position of\_Char\_in\_Alphabet$$
>
> $$\psi_{Char} = \frac{1}{Current\_Char\_Weight - Previous\_Char\_Weight} + \frac{\pi}{100}$$
>
> $$Char\_Offset = \left| \psi_{first\_character} \times 10 \right|$$
>
> $$Char\_Total\_Value = Char\_Offset \times Char\_Weight$$
>
> **Step 1.2** Calculate the Block _Value using following equation
>
> $$Block\_Value = \left(\sum_{i=1}^{n} Char_i\_Total\_Value \right) - Length\_of\_Block$$
>
> **Step 1.3** Perform the following operation on each block depending on the block value.
>
> $$if (0 \leq Block\_Value < 100) then$$
>
> **Reverse the block**
>
> $$if (100 \leq Block\_Value < 150) then$$
>
> **Block is circular right shifted by** $\dfrac{(Length\_of\_Block)}{2}$
>
> $$if (150 \leq Block\_Value < 200) then$$
>
> **Block is circular left shifted by** $\dfrac{(Length\_of\_Block)}{2}$

Step 2. Perform XOR operation between level 2 encoded plain text and SA based keystream to generate the level 3encoded text.

Step 3. Finally perform xor operation between MLP based session key and level 3 encoded plain text and then transfer to the receiver.

Step 4. Perform the Automata based encryption to get level1 encoded SA based keystream. This technique ensures that if intruders intercept the key of the keystream then also values of the key not be known to the intruders because of the Automata based encoding.

> **Step 4.1** Break the SA based keystream into 3 bit binary code and use the following table to represent value of each code and their octal, gray code representation and state number.

Table 3. Octal, gray code representation and state number table

Binary (3 bit)	Octal	Gray Code	State Number
000	0	000	q0
001	1	001	q1
010	2	011	q2
011	3	010	q3
100	4	111	q4
101	5	110	q5
110	6	100	q6
111	7	101	q7

For example if SA based key stream is in the form 100 110 001 011 101 then state representation will be Q= set of finite state i.e. {q4, q6, q1, q3, q5} where q4 is the initial state and q5 is the final state. State transition operator δ is use to denote transition from one state to another and Σ denotes alphabet set. Then state transition is done using following representations:

$\delta(q4, tranmsition\_variable) = q6$ i.e. $(q4 \longrightarrow q6)$

$\delta(q6, tranmsition\_variable) = q1$ i.e. $(q6 \longrightarrow q1)$

$\delta(q1, tranmsition\_variable) = q3$ i.e. $(q1 \longrightarrow q3)$

$\delta(q3, tranmsition\_variable) = q5$ i.e. $(q3 \longrightarrow q5)$

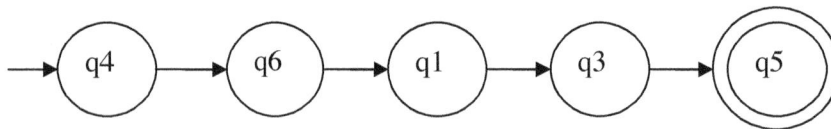

Step 4.2   Perform the following operation

$$
\begin{aligned}
&\{ \max(\ q_4, q_6\ ) in\_octal\_form \times \max(\ q_4, q_6\ ) in\_gray\_code\ \} + \\
&\{ \max(\ q_6, q_1\ ) in\_octal\_form \times \max(\ q_6, q_1\ ) in\_gray\_code\ \} + \\
&\{ \max(\ q_1, q_3\ ) in\_octal\_form \times \max(\ q_1, q_3\ ) in\_gray\_code\ \} + \\
&\{ \max(\ q_3, q_5\ ) in\_octal\_form \times \max(\ q_3, q_5\ ) in\_gray\_code\ \}
\end{aligned}
$$

Step 4.3   Result of this operation get converted in binary form and then get XOR-ed with SA based keystream value in reverse order to produce level1 encoded keystream.

Step 5. Perform xoring between MLP based session key and level 1 encoded key stream to produce level 2 based key and transmitted to the receiver.

5. DECRYPTION ALGORITHM

Step 1. Receive the cipher text and level2 based encoded keystream from the sender.

Step 2. Generate level1 based encoded keystream by performing xoring between MLP based session key and level 2 encoded key stream.

Step 3. Generate SA based keystream by performing Automata based decryption on level1 encoded keystream.

Step 3.1   Break the level1 encoded keystream into 3 bit binary code and use the following table to represent value of each code and their octal, gray code representation and state number.

Table 4. Octal, gray code representation and state number table

Binary (3 bit)	Octal	Gray Code	State Number
000	0	000	q0
001	1	001	q1
010	2	011	q2
011	3	010	q3
100	4	111	q4
101	5	110	q5
110	6	100	q6
111	7	101	q7

For example if level1 encoded key stream is in the form 100 110 001 011 101 then state representation will be Q= set of finite state i.e. {q4, q6, q1, q3, q5} where q4 is the initial state and q5 is the final state. State transition operator δ is use to denote transition from one state to another and Σ denotes alphabet set. Then state transition is done using following representations:

$\delta(q4, tranmsition\_variable) = q6$ i.e. (q4 → q6)

$\delta(q6, tranmsition\_variable) = q1$ i.e. (q6 → q1)

$\delta(q1, tranmsition\_variable) = q3$ i.e. (q1 → q3)

$\delta(q3, tranmsition\_variable) = q5$ i.e. (q3 → q5)

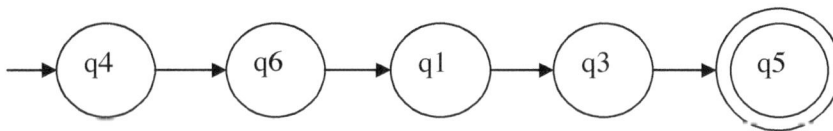

Step 3.2   Now, perform the following operation

$$\{ max(q_4, q_6) in\_octal\_form \times max(q_4, q_6) in\_gray\_code \} +$$
$$\{ max(q_6, q_1) in\_octal\_form \times max(q_6, q_1) in\_gray\_code \} +$$
$$\{ max(q_1, q_3) in\_octal\_form \times max(q_1, q_3) in\_gray\_code \} +$$
$$\{ max(q_3, q_5) in\_octal\_form \times max(q_3, q_5) in\_gray\_code \}$$

Step 3.3   Result of this operation get converted in binary form and then get XOR-ed with level1 encoded keystream value in reverse order to produce SA based keystream.

Step 4. Perform xor operation between MLP encoded text and MLP based session key to generate level 3 encoded text.

Step 5. Perform xor operation between level 3 encoded text and SA based key to generate level 2 encoded text.

Step 6. Generate level1 based encoded plain text by performing Comparison based decryption on level2 encoded plain text

Step 6.1   For the characters in the block perform following calculations. For the very first character in the block weight value of the previous character will be 0.

$$Char\_Weight = Positionof\_Char\_in\_Alphabet$$

$$\psi_{Char} = \frac{1}{Current\_Char\_Weight - Previous\_Char\_Weight} + \frac{\pi}{100}$$

$$Char\_Offset = \left| \psi_{first\_character} \times 10 \right|$$

$$Char\_Total\_Value = Char\_Offset \times Char\_Weight$$

Step 6.2   Calculate the Block _Value using following equation

$$Block\_Value = \left(\sum_{i=1}^{n} Char_i\_Total\_Value \right) - Length\_of\_Block$$

Step 6.3 Perform the following operation on each block depending on the block value.

$$if\,(0 \leq Block\_Value < 100)then$$

Reverse the block

$$if\,(100 \leq Block\_Value < 150)then$$

Block is circular right shifted by $\dfrac{(Length\_of\_Block)}{2}$

$$if\,(150 \leq Block\_Value < 200)then$$

Block is circular left shifted by $\dfrac{(Length\_of\_Block)}{2}$

Step 7.   Generate plain text by performing metamorphosed character code based decryption on level1 encoded plain text

6. SIMULATED ANNEALING (SA) BASED KEY GENERATION

Simulated Annealing (SA) algorithm which is based on the analogy between the annealing of solids and the problem of solving combinatorial optimization problems SA provides local search technique that helps to escape from local optima. In this scheme SA is used to generate key after satisfying some desired features such as good statistical properties, long period, large linear complexity, and highly order degree of correlation immunity.

In proposed SA based key generation technique Section A deals with initialization procedure and that of section B describes the fitness calculation, section C describes cooling mechanism and finally section D deals with SA based key generation algorithm.

Initialization Procedure

At the preliminary state of SA, each sequence is represented as a binary string of an equal number of 0's and 1's of a given length. So, each generated keystream (solution) coded as a binary string.

Fitness Calculation

For each sequence fitness value is calculated examining the keystream. Using the following sequence of steps fitness is calculated:

Frequency test of 0's & 1's: Proportion of 0's & 1's in the total sequence are being checked using eq. $Frequency\_Fault = \left| \psi_0 - \psi_1 \right|$ Ψ_0: No. of 0's in total sequence. Ψ_1: No. of 1's in total sequence.

Binary Derivative Test: Binary Dervative test is applied to the sequence by taking the overlapping 2 tupels in the original bit stream.

Table 5. Binary xor Operation Table

First Bit	Second Bit	Result
0	0	0
0	1	1
1	0	1
1	1	0

Here, an equal proportion of 1's and 0's in the new bit stream is checked by eq.

$$Binary\_DerivativeFault = \left| \left(C\_\psi_0 - C\_\psi_1 \right) - 1 \right|$$

Where, C_Ψ_0: Count no. of 0's in new bit stream. C_Ψ_1: Count no. of 1's in new bit stream.

Change Point Test: In this test a check point is created for observing maximum difference between, the proportions of 1's including the check point and the proportions of 1's after the check point using following eq.

$$D[C_p] = \Psi * K[C_p] - C_{p*} K[\Psi] \qquad \rho_r = \exp\left(-2M^2 / \psi * K[\psi] * \left(\psi - K[\psi] \right) \right)$$

$$Change\_Po\,int\_Fault = \rho_r$$

Ψ: Total no. of bit in the stream. $K[\Psi]$: Total no. of 1's in the bit stream. C_p: Change Point. $K[C_p]$: Total no. of 1's to bit C_p (Change Point). $D[C_p]$: Difference respect to the Change point. M: MAX (ABS ($D[C_p]$)), for $C_{p=1..}\Psi$.

ρ_r Probability of statistics that smaller value of ρ_r more significant the result. Finally, fault of every test is summed up for calculating fitness function. Using following eq. fitness is calculated.

$$Fitness\_Function = \frac{1}{1 + Fault}$$

Cooling Procedure

To determining the cooling schedule in case of optimization problem in annealing process requires some parameters for initial value of control parameter, decrement function of control parameter, length of individual distance parameter, stopping criteria

SA based Key Generation Algorithm

Step 1*: Set T as a starting temperature.Set Itr (no. of iteration) =0. Set Inner_Cycle=0, Set Finish=False.*
Step 2: *Randomly generate sequence of binary bits and calculate the fitness function.*
Where, $F_{current}$= fitness function of current state.
Step 3: *While Finish=False do (a) to (d) Repeat inner loop iteration n times.*
(a) Generate next sequence using swapping & flipping operators.
(b) Evaluate the fitness value F_{new} of the new sequence.
(c) Evaluate the energy changes in fitness function. $\Delta E = F_{new} - F_{current}$

(d) If $F_{new} > F_{current}$ then new solution has a better fitness than the current one so accept the move. $F_{current} = F_{new}$.

Else evaluate $\rho_r = \exp\left(\dfrac{-\Delta E}{T}\right)$

If $\rho_r > $ Threshold value then accept the move.

Else Reject. In the recent inner loop if no move is accepted then   Inner_Cycle= Inner_Cycle+1;   T=T*α;   Itr=Itr+1;   If (Inner_Cycle > Max_Cycle) or (Itr>Max_Itr) then Finish=True

Step 4:   *$F_{current}$ becomes the final solution.*
Now, the final solution i.e. stream of 0's & 1's becomes the SA based sub key for 2nd level encryption.

7. STRUCTURE OF MULTILAYER PERCEPTRON

In multilayer perceptron synchronization scheme secret session key is not physically get exchanged over public insecure channel. At end of neural weight synchronization strategy of both parties' generates identical weight vectors and activated hidden layer outputs for both the parties become identical. This identical output of hidden layer for both parties can be use as one time secret session key for secured data exchange. A multilayer perceptron synaptic simulated weight based undisclosed key generation is carried out between recipient and sender. Figure5 shows multilayer perceptron based synaptic simulation system. Sender and receivers multilayer perceptron select same single hidden layer among multiple hidden layers for a particular session. For that session all other hidden layers goes in deactivated mode means hidden (processing) units of other layers do nothing with the incoming input. Either synchronized identical weight vector of sender and receivers' input layer, activated hidden layer and output layer becomes session key or session key can be form using identical output of hidden units of activated hidden layer.

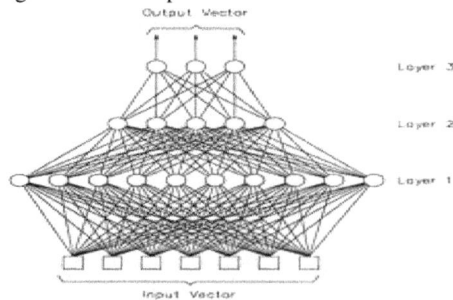

Figure 5. A Multilayer Perceptron with 3 Hidden Layers

Sender and receiver multilayer perceptron in each session acts as a single layer network with dynamically chosen one activated hidden layer and K no. of hidden neurons, N no. of input neurons having binary input vector, $x_{ij} \in \{-1,+1\}$, discrete weights, are generated from input to output, are lies between -L and +L, $w_{ij} \in \{-L,-L+1,...+L\}$. Where i = 1,…,K denotes the ith hidden unit of the perceptron and j = 1,…,N the elements of the vector and one output neuron. Output of the hidden units is calculated by the weighted sum over the current input values . So, the state of the each hidden neurons is expressed using (eq.1)

$$h_i = \frac{1}{\sqrt{N}} w_i x_i = \frac{1}{\sqrt{N}} \sum_{j=1}^{N} w_{i,j} x_{i,j} \tag{1}$$

Output of the ith hidden unit is defined as

$$\sigma_i = \mathrm{sgn}(\, h_i)$$

(2)

But in case of $h_i = 0$ then $\sigma_i = -1$ to produce a binary output. Hence a, $\sigma_i = +1$, if the weighted sum over its inputs is positive, or else it is inactive, $\sigma_i = -1$. The total output of a perceptron is the product of the hidden units expressed in (eq. 2)

$$\tau = \prod_{i=1}^{K} \sigma_i$$

(3)

The learning mechanism proceeds as follows ([6, 7]):

1.   If the output bits are different, $\tau^A \neq \tau^B$, nothing is changed.
2.   2. If $\tau^A = \tau^B = \tau$, only the weights of the hidden units with $\sigma_k^{A/B} = \tau^{A/B}$ will be updated.
3. The weight vector of this hidden unit is adjusted using any of the following learning rules:

Anti-Hebbian:

$$W_k^{A/B} = W_k^{A/B} - \tau^{A/B} x_k \Theta(\sigma_k \tau^{A/B})(\tau^A \tau^B)$$

(4)

Hebbian :

$$W_k^{A/B} = W_k^{A/B} + \tau^{A/B} x_k \Theta(\sigma_k \tau^{A/B})(\tau^A \tau^B)$$

(5)

Random walk:

$$W_k^{A/B} = W_k^{A/B} + x_k \Theta(\sigma_k \tau^{A/B})(\tau^A \tau^B)$$

(6)

During step (2), if there is at least one common hidden unit with $\sigma_k = \tau$ in the two networks, then there are 3 possibilities that characterize the behaviour of the hidden nodes:

1. An attractive move: if hidden units at similar k positions have equal output bits, $\sigma_k^A = \sigma_k^B = \tau^{A/B}$

2. A repulsive move: if hidden units at similar k positions have unequal output bits, $\sigma_k^A \neq \sigma_k^B$

3. No move: when $\sigma_k^A = \sigma_k^B \neq \tau^{A/B}$

The distance between hidden units can be defined by their mutual overlap, ρ_k,

$$\rho_k = \frac{w_k^A\, w_k^B}{\sqrt{w_k^A\, w_k^A}\, \sqrt{w_k^B\, w_k^B}}$$

(7)

where $0 < \rho_k < 1$, with $\rho_k = 0$ at the start of learning and $\rho_k = 1$ when synchronization occurs with the two hidden units having a common weight vector.

7.1 Multilayer Perceptron Simulation Algorithm

Input: - Random weights, input vectors for both multilayer perceptrons.
Output: - Secret key through synchronization of input and output neurons as vectors.
Method:-

Step 1. *Initialization of random weight values of synaptic links between input layer and randomly selected activated hidden layer.*

$$Where, \quad w_{ij} \in \{-L, L+1, ... H_j\} \tag{8}$$

Step 2. *Repeat step 3 to 6 until the full synchronization is achieved, using Hebbian-learning rules.*

$$w_{i,j}^+ = g\left(w_{i,j} + x_{i,j}\tau\Theta(\sigma_i\tau)\Theta(\tau^A\tau^B)\right) \tag{9}$$

Step 3. *Generate random input vector X. Inputs are generated by a third party or one of the communicating parties.*

Step 4. *Compute the values of the activated hidden neurons of activated hidden layer using (eq. 10)*

$$h_i = \frac{1}{\sqrt{N}} w_i x_i = \frac{1}{\sqrt{N}} \sum_{j=1}^{N} w_{i,j} x_{i,j} \tag{10}$$

Step 5. *Compute the value of the output neuron using*

$$\tau = \prod_{i=1}^{K} \sigma_i \tag{11}$$

Compare the output values of both multilayer perceptron by exchanging the system outputs.
if Output (A) ≠ Output (B), Go to step 3
else if Output (A) = Output (B) then one of the suitable learning rule is applied
only the hidden units are trained which have an output bit identical to the common output.

Update the weights only if the final output values of the perceptron are equivalent. When synchronization is finally achieved, the synaptic weights are identical for both the system.

7.2 Multilayer Perceptron Learning rule

At the beginning of the synchronization process multilayer perceptron of A and B start with uncorrelated weight vectors $w_i^{A/B}$. For each time step K, public input vectors are generated randomly and the corresponding output bits $\tau^{A/B}$ are calculated. Afterwards A and B communicate their output bits to each other. If they disagree, $\tau^A \neq \tau^B$, the weights are not changed. Otherwise learning rules suitable for synchronization is applied. In the case of the Hebbian learning rule [10] both neural networks learn from each other.

$$w_{i,j}^+ = g\left(w_{i,j} + x_{i,j}\tau\Theta(\sigma_i\tau)\Theta(\tau^A\tau^B)\right) \tag{12}$$

The learning rules used for synchronizing multilayer perceptron share a common structure. That is why they can be described by a single (eq. 4)

$$w_{i,j}^+ = g\left(w_{i,j} + f(\sigma_i, \tau^A, \tau^B)x_{i,j}\right) \tag{13}$$

with a function $f(\sigma_i, \tau^A, \tau^B)$, which can take the values -1, 0, or +1. In the case of bidirectional interaction it is given by

$$f\left(\sigma_i, \tau^A, \tau^B\right) = \Theta\left(\sigma\tau^A\right)\Theta\left(\tau^A\tau^B\right)\begin{cases}\sigma & \textit{Hebbian learning}\\ -\sigma & \textit{anti-Hebbian learning}\\ 1 & \textit{Random walk learning}\end{cases} \tag{14}$$

The common part $\Theta\left(\sigma\tau^A\right)\Theta\left(\tau^A\tau^B\right)$ of $f\left(\sigma_i, \tau^A, \tau^B\right)$ controls, when the weight vector of a hidden unit is adjusted. Because it is responsible for the occurrence of attractive and repulsive steps [6].

The equation consists of two parts:

1. $\Theta\left(\sigma\tau^A\right)\Theta\left(\tau^A\tau^B\right)$: This part is common between the three learning rules and it is responsible for the attractive and repulsive effect and controls when the weight vectors of a hidden unit is updated. Therefore, all three learning rules have similar effect on the overlap.

2. $(\sigma, -\sigma, 1)$: This part differs among the three learning rules and it is responsible for the direction of the weights movement in the space. Therefore, it changes the distribution of the weights in the case of Hebbian and anti-Hebbian learning. For the Hebbian rule, A's ad B's multilayer perceptron learn their own output and the weights are pushed towards the boundaries at $-L$ and $+L$. In contrast, by using the anti- Hebbian rule, A's and B's multilayer perceptron learn the opposite of their own outputs. Consequently, the weights are pulled from the boundaries $\pm L$. The random walk rule is the only rule that does not affect the weight distribution so they stay uniformly distributed. In fact, at large values of N, both Hebbian and anti-Hebbian rules do not affect the weight distribution. Therefore, the proposed algorithm is restricted to use either random walk learning rule or Hebian or anti-Hebbian learning rules only at large values of N. The random walk learning rule is chosen since it does not affect the weights distribution regardless of the value of N.

7.5 Hidden Layer as a Secret Session Key

At end of full weight synchronization process, weight vectors between input layer and activated hidden layer of both multilayer perceptron systems become identical. Activated hidden layer's output of source multilayer perceptron is used to construct the secret session key. This sessionkey is not get transmitted over public channel because receiver multilayer perceptron has same identical activated hidden layer's output. Compute the values of the each hidden unit by

$$\sigma_i = \text{sgn}\left(\sum_{j=1}^{N} w_{ij}x_{ij}\right) \quad \text{sgn}(x) = \begin{cases}-1 & \textit{if } x < 0,\\ 0 & \textit{if } x = 0,\\ 1 & \textit{if } x > 0.\end{cases} \tag{15}$$

For example consider 8 hidden units of activated hidden layer having absolute value (1, 0, 0, 1, 0, 1, 0, 1) becomes an 8 bit block. This 10010101 become a secret session key for a particular session and cascaded xored with recursive replacement encrypted text. Now final session key based encrypted text is transmitted to the receiver end. Receiver has the identical session key i.e. the output of the hidden units of activated hidden layer of receiver. This session key used to get the recursive replacement encrypted text from the final cipher text. In the next session both the machines started tuning again to produce another session key.

Identical weight vector derived from synaptic link between input and activated hidden layer of both multilayer perceptron can also becomes secret session key for a particular session after full weight synchronization is achieved.

8. THE GSMLPSA TECHNIQUE

Our proposed group key swap over technique offers two novel procedures for exchanging group key among different multilayer perceptron. Both procedures are based on the structure of complete binary tree. In the multilayer perceptron group key exchange algorithm, N multilayer perceptrons need to synchronize together and they are represented by an M number of leaves of a complete binary tree where M is defined as $M = 2^{(\log_2 N)}$.

Complete binary tree based proposed procedures are
   a) Complete Binary Tree with Vote (CBTV)
   b) Complete Binary Tree with Exchange (CBTE)
Both procedures have the same end results but their implementations are different.

8.1. Synchronization using Complete Binary Tree with Vote (CBTV)

In the CBTV method, the N multilayer perceptrons are represented by the M leaves. For every step j (starting at $j = 1$) of the algorithm the binary tree is divided into $\dfrac{M}{2^j}$ subtrees each with 2^j leaves. Then each pair of leaves sharing the same parent involved in mutual learning. Next, j is incremented and in each subtree, a node is nominated and the mutual learning algorithm is executed by the nominated nodes and the rest if the nodes follow. When the algorithm reaches the root, then it terminates and hence, all the multilayer perceptrons are synchronized and share the same weight vectors.

Algorithm 1 –The CBTV method
1: loop { for j = $\log_2 M$; $j \geq 0$; j --}
2: Nominate a leader in each left and right subtree in level j
3: Apply Mutual learning between nominated leaders
4: Every leader sends the τ bits to its group.
5: end loop

8.2. Synchronization using Complete Binary Tree with Exchange (CBTE)

In the CBTE method, the mutual learning algorithm is take place between every two parties having the same parent in the binary tree structure. Let the *max depth* is the depth of the complete binary tree and *cur depth* is the current depth where the algorithm is functioning. Starting from a *cur depth* = *max depth*−1, apply the mutual learning algorithm between each pair of leaves having the same parent. Following the synchronization, one level up is marked
(*cur depth* = *cur depth* − 1) and a exchange method is applied between the right leaves of both right and left branches for all subtrees in that *cur depth*. Once the *cur depth* becomes equal to zero, all leaves will be synchronized together. For sake of simplicity the group of parties will be represented as vector with indices $\{0, 1, \ldots, M-1\}$ Fig.6 shows the scenario of synchronization. Fig.6a shows the preliminary configuration of unsynchronized parties. In Fig.6b, pairs of parties are synchronized together, $\{(0, 1), (2, 3), (4, 5), (6, 7)\}$. Then, the exchange operation is performed, $\{(0, 2), (1, 3), (4, 6), (6, 7)\}$, and the mutual learning is applied again. This results in

synchronization of two groups each with four parties, {(0, 1, 2, 3), (4, 5, 6, 7)}, as shown in Fig.6c. After that, the exchange operation is applied again and the vector takes the form {(0, 4), (1, 5), (2, 6), (3, 7)}. The algorithm terminates when pairs in the new vector apply mutual learning that produces full synchronization between all parties (Fig.6d). The CBTV method needs to transmit the data between the nominated nodes to other nodes in order to be followed. On the other hand, the CBTE algorithm applies the mutual learning algorithm between each pair of nodes separately.

At the same time as the proposed key exchange protocol is scalable; it remains susceptible to active attacks. An attacker can take part in the protocol and synchronize with the group and finally obtain the shared key which endangers the secret communication between the group later. As a result, it is compulsory to build up an authenticated key exchange protocol to permit only certified users to get hold of the mutual secret.

Algorithm 2 –The CBTE method

Require: l, m variables.
1: loop { for j = 0; $j \le log_2 M$; j ++}
2: $m = 2j$
3: A node participates just the once per iteration.
4: Apply mutual learning between nodes l, $l + m$
5: end loop

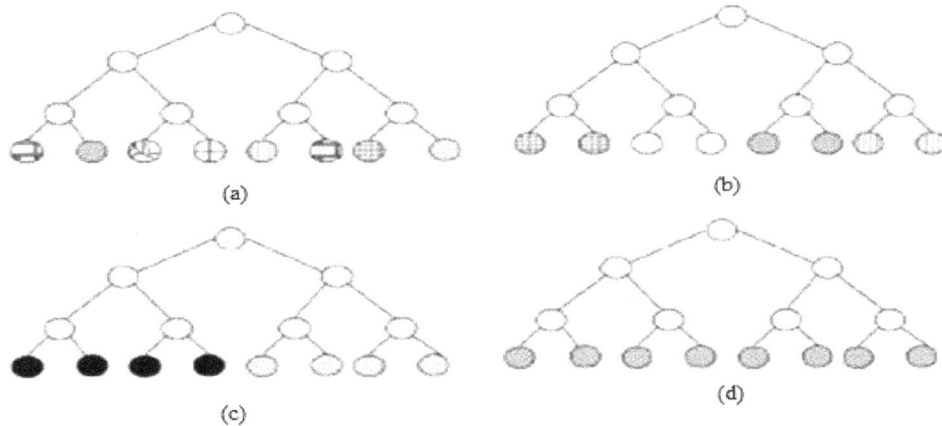

Figure 6. (a) Shows the preliminary configuration of unsynchronized parties.
(b) Pairs of parties are synchronized together, {(0, 1), (2, 3), (4, 5), (6, 7)}.
(c) Synchronization of two groups each with four parties, {(0, 1, 2, 3), (4, 5, 6, 7)}.
(d) After that, the exchange operation is applied again and the vector takes the form   {(0, 4), (1, 5), (2, 6), (3, 7)}.

9. GSMLPSA CERTIFICATE GENERATION

While proposed GSMLPSA method offers rapid key exchange between groups of users, it is susceptible to malicious attacks where an challenger can participate in the protocol and hence obtains the group top secret key. So, the group requires an authentication certificate to safeguard it against such types of attacks. In order to construct an authentication certificate, GSMLPSA assumes that the group obtains a secret password which can be used to authenticate the exchange protocol. This password can be mapped to multilayer perceptron guided cryptographic public parameter which can be used as an initial seed for a random number generator which encrypts the output bits τ in a fashion similar to that was proposed in [12, 13]. Assume a random number generator (RNG), R_i, $R_i^p = \mathrm{Re}\,m((a * R_{i-1}^p \_ c), m)$ with the set $\lambda = \{a, c, m\}$ being the RNG parameters.

Algorithm 3 –GSMLPSA Password Authentication Scheme

Require: n is the number of iterations needed to synchronize between two parties and a nonlinear function F.

loop {for each pair}

  Generate random number R_0 and publicly exchange between each pair.

  Compute $R_1 = R_0 \oplus password$

  loop {for n}

    Generate random number R_i

    if $F(R_i^A > 0)$ then

      $\tau_{sent}^A = -\tau_{compute}^A$

    end if

    if $F(R_i^B, R_i^B) > 0$ then

      $\tau_{used}^A = -\tau_{sent}^A$

    end if

  end loop

end loop

---

10. COMPLEXITY ANALYSIS

The complexity of the Synchronization technique will be O(L), which can be computed using following three steps.

Step 1. To generate a MLP guided key of length N needs O(N) Computational steps. The average synchronization time is almost independent of the size N of the networks, at least up to N=1000.Asymptotically one expects an increase like O (log N).

Step 2. Complexity of the encryption technique is O(L).

  Step 2.1. Recursive replacement of bits using prime nonprime recognition encryption process takes O(L).

  Step 2.2. MLP based encryption technique takes O(L) amount of time.

Step 3. Complexity of the decryption technique is O(L).

  Step 3.1. In MLP based decryption technique, complexity to convert final cipher text into recursive replacement cipher text T takes O(L).

Key exchange algorithm has complexity of logarithmic proportional to the number of the parties need to synchronize together. Because key exchange protocols works on a structure of a complete binary tree. Algorithm works form leaf level to the root i.e. the height of a complete binary tree which is O (log N).

11. EXPERIMENT RESULTS

In this section, CBTV is applied between group of parties and some simulation results are presented. For simplicity, the number of communicating parties is taken to be four. i.e., ($M = 4$). Assuming four parties A, B, C and D need to share a common key so they apply the CBTV algorithm. As shown in Fig.7 curve 1, A and B apply the ordinary mutual learning algorithm till they synchronize. At the same time both C and D do the same as shown in curve 2. Then the swapping mechanism is applied and hence, A and C apply the ordinary mutual learning algorithm and the same scenario repeats for B and D. It is evident that curves 3 and 4 are identical which indicates that the four parties have synchronized at common weight vectors. If another party requires to share a key with previously N synchronized parties, it does not need to repeat the entire algorithm again. Instead, the N synchronized parties are dealt with as a single partner and the mutual learning algorithm is applied between an elected party of the N partners

and the new party. Then the other $(N - 1)$ parties apply the learning rules without sending their output bits over the public channel.

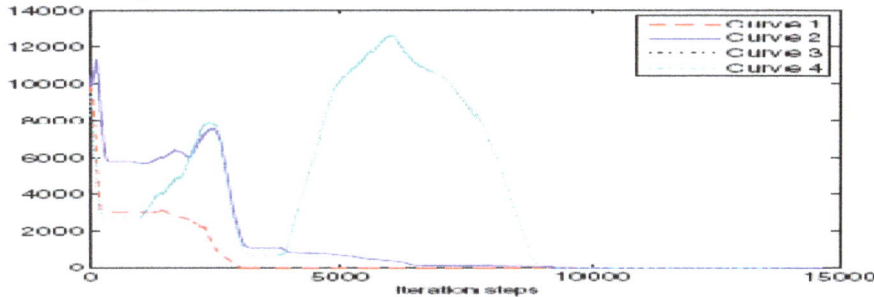

Figure 7. Synchronization between 4 Multilayer Perceptrons.

In this section the results of implementation of the proposed GSMLPSA encryption/decryption technique has been presented in terms of encryption decryption time, Chi-Square test, source file size vs. encryption time along with source file size vs. encrypted file size. The results are also compared with existing RSA [1] technique, existing ANNRBLC [8] and NNSKECC [9].

Table 6. Encryption / decryption time vs. File size

Encryption Time (s)			Decryption Time (s)		
Source Size (bytes)	GSMLPSA	NNSKECC [9]	Encrypted Size (bytes)	GSMLPSA	NNSKECC [9]
18432	6. 42	7.85	18432	6.99	7.81
23044	9. 23	10.32	23040	9.27	9.92
35425	14. 62	15.21	35425	14. 47	14.93
36242	14. 72	15.34	36242	15. 19	15.24
59398	25. 11	25.49	59398	24. 34	24.95

Table 6 shows encryption and decryption time with respect to the source and encrypted size respectively. It is also observed the alternation of the size on encryption.

Table 7 shows Chi-Square value for different source stream size after applying different encryption algorithms. It is seen that the Chi-Square value of GSMLPSA is better compared to the algorithm ANNRBLC [8] and also better than SATMLP [17] algorithm because of Metamorphosed guided Automata based comparison encryption rather than simple triangularized encryption in case of SATMLP [17].

Table 7. Source size vs. Chi-Square value

Stream Size (bytes)	Chi-Square value (TDES) [1]	Chi-Square value (ANNRBLC) [8]	Chi-Square value in (SATMLP) [17]	Chi-Square value Proposed GSMLPSA
1500	1228.5803	2471.0724	2627.7534	2718.2517
2500	2948.2285	5645.3462	5719.8522	5935.0471
3000	3679.0432	6757.8211	6739.73621	6938.7203
3250	4228.2119	6994.6198	7009.2813	7114.0649
3500	4242.9165	10572.4673	11624.2315	11729.8530

Figure 8 shows graphical representation of table 7.

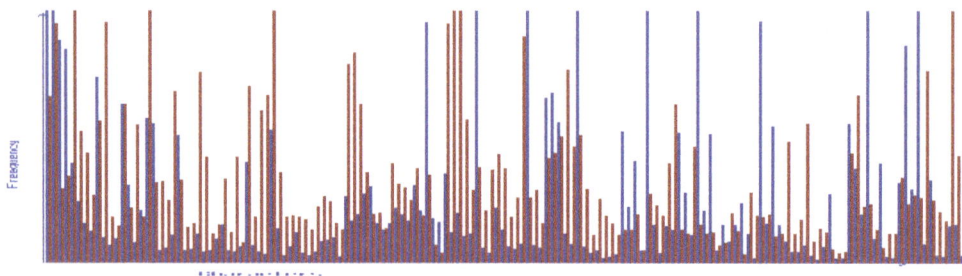

Figure 8. Chi-Square value against stream size

Table 8 shows total number of iteration needed and number of data being transferred for GSMLPSA key generation process with different numbers of input(N) and activated hidden(H) neurons and varying synaptic depth(L).

Table 8. Data Exchanged and No. of Iterations For Different Parameters Value

No. of Input Neurons(N)	No. of Activated Hidden Neurons(K)	Synaptic Weight (L)	Total No. of Iterations	Data Exchanged (Kb)
5	15	3	624	48
30	4	4	848	102
25	5	3	241	30
20	10	3	1390	276
8	15	4	2390	289

12. ANALYSIS OF RESULTS

From results obtained it is clear that the technique will achieve optimal performances. Encryption time and decryption time varies almost linearly with respect to the block size. For the algorithm presented, Chi-Square value is very high compared to some existing algorithms. A user input key has to transmit over the public channel all the way to the receiver for performing the decryption procedure. So there is a likelihood of attack at the time of key exchange. To defeat this insecure secret key generation technique a neural network based secret key generation technique has been devised. The security issue of existing algorithm can be improved by using GSMLPSA secret session key generation technique. In this case, the two partners A and B do not have to share a common secret but use their indistinguishable weights or output of activated hidden layer as a secret key needed for encryption. The fundamental conception of GSMLPSA based key exchange protocol focuses mostly on two key attributes of GSMLPSA. Firstly, two nodes coupled over a public channel will synchronize even though each individual network exhibits disorganized behaviour. Secondly, an outside network, even if identical to the two communicating networks, will find it exceptionally difficult to synchronize with those parties, those parties are communicating over a public network. An attacker E who knows all the particulars of the algorithm and records through this channel finds it thorny to synchronize with the parties, and hence to calculate the common secret key. Synchronization by mutual learning (A and B) is much quicker than learning by listening (E) [10]. For usual cryptographic systems, we can improve the safety of the protocol by increasing of the key length. In the case of GSMLPSA, we improved it by increasing the synaptic depth L of the neural networks. For a brute force attack using K hidden neurons, K*N input neurons and boundary of weights L, gives (2L+1)KN possibilities. For example, the configuration K = 3, L = 3 and N = 100 gives us 3*10253 key possibilities, making the attack unfeasible with today's computer power. E could start from all of the (2L+1)3N initial weight vectors and calculate the ones which are consistent with the input/output sequence. It has been shown, that all of these initial states move towards the same final weight vector, the key is unique. This is not true for simple perceptron the most unbeaten cryptanalysis has two supplementary ingredients first; a group of attacker is used. Second, E makes extra training steps when A and B are quiet [10]-[12]. So increasing synaptic depth L of the GSMLPSA we can make our GSMLPSA safe.

13. SECURITY ISSUE

The main difference between the partners and the attacker in GSMLPSA is that A and B are able to influence each other by communicating their output bits τ^A & τ^B while E can only listen to these messages. Of course, A and B use their advantage to select suitable input vectors for adjusting the weights which finally leads to different synchronization times for partners and attackers. However, there are more effects, which show that the two-way communication between A and B makes attacking the GSMLPSA protocol more difficult than simple learning of examples. These confirm that the security of GSMLPSA key generation is based on the bidirectional interaction of the partners. Each partener uses a seperate, but identical pseudo random number generator. As these devices are initialized with a secret seed state shared by A and B. They produce exactly the same sequence of input bits. Whereas attacker does not know this secret seed state. By increasing synaptic depth average synchronize time will be increased by polynomial time. But success probability of attacker will be drop exponentially Synchonization by mutual learning is much faster than learning by adopting to example generated by other network. Unidirectional learning and bidirectional synchronization. As E can't influence A and B at the time they stop transmit due to synchrnization. Only one weight get changed where, = T. So, difficult to find $\sigma_i$weight for attacker to know the actual weight without knowing internal representation it has to guess.

14. FUTURE SCOPE & CONCLUSION

This paper presented a novel approach for group key exchange. GSMLPSA algorithms are proposed as extensions to the ordinary mutual learning algorithm. Also it has been shown that the complexity of the algorithms is logarithmic proportional to the number of the parties need to synchronize together. This algorithm can be used in many applications such as video and voice conferences. This technique enhances the security features of the key exchange algorithm by increasing of the synaptic depth L of the GSMLPSA. Here two partners A and B do not have to exchange a common secret key over a public channel but use their indistinguishable weights or outputs of the activated hidden layer as a secret key needed for encryption or decryption. So likelihood of attack proposed technique is much lesser than the simple key exchange algorithm.

Future scope of this technique is that this GSMLPSA model can be used in wireless communication and also in key distribution mechanism.

ACKNOWLEDGEMENTS

The author expresses deep sense of gratitude to the Department of Science & Technology (DST) , Govt. of India, for financial assistance through INSPIRE Fellowship leading for a PhD work under which this work has been carried out, at the department of Computer Science & Engineering, University of Kalyani.

REFERENCES

[1]   Atul Kahate, Cryptography and Network Security, 2003, Tata McGraw-Hill publishing Company Limited, Eighth reprint 2006.

[2]   Sarkar Arindam, Mandal J. K, "Artificial Neural Network Guided Secured Communication Techniques: A Practical Approach" LAP Lambert Academic Publishing (2012-06-04), ISBN: 978-3-659-11991-0, 2012

[3]   Sarkar Arindam, Karforma S, Mandal J. K, "Object Oriented Modeling of IDEA using GA based Efficient Key Generation for E-Governance Security (OOMIG) ", International Journal of Distributed and Parallel Systems (IJDPS) Vol.3, No.2, March 2012, DOI : 10.5121/ijdps.2012.3215, ISSN : 0976 - 9757 [Online] ; 2229 - 3957 [Print]. Indexed by: EBSCO, DOAJ, NASA, Google Scholar, INSPEC and WorldCat, 2011.

[4]   Mandal J. K., Sarkar Arindam, "Neural Session Key based Traingularized Encryption for Online Wireless Communication (NSKTE)", 2nd National Conference on Computing and Systems, (NaCCS 2012), March 15-16, 2012, Department of Computer Science, The University of Burdwan, Golapbag North, Burdwan –713104, West Bengal, India. ISBN 978- 93-808131-8-9, 2012.

[5]   Mandal J. K., Sarkar Arindam, "Neural Weight Session Key based Encryption for Online Wireless Communication (NWSKE)", Research and Higher Education in Computer Science and Information Technology, (RHECSIT- 2012) ,February 21-22, 2012, Department of Computer Science, Sammilani Mahavidyalaya, Kolkata , West Bengal, India. ISBN 978-81- 923820-0-5,2012

[6]   Mandal J. K., Sarkar Arindam, "An Adaptive Genetic Key Based Neural Encryption For Online Wireless Communication (AGKNE)", International Conference on Recent Trends In Information Systems (RETIS 2011) BY IEEE, 21-23 December 2011, Jadavpur University, Kolkata, India. ISBN 978-1-4577-0791-9, 2011

[7]   Mandal J. K., Sarkar Arindam, "An Adaptive Neural Network Guided Secret Key Based Encryption Through Recursive Positional Modulo-2 Substitution For Online Wireless Communication (ANNRPMS)", International Conference on Recent Trends In Information Technology (ICRTIT 2011) BY IEEE, 3-5 June 2011, Madras Institute of Technology, Anna University, Chennai, Tamil Nadu, India. 978-1-4577-0590-8/11, 2011

[8]   Mandal J. K., Sarkar Arindam, "An Adaptive Neural Network Guided Random Block Length Based Cryptosystem (ANNRBLC)", 2nd International Conference on Wireless Communications, Vehicular Technology, Information Theory And Aerospace & Electronic System Technology" (Wireless Vitae 2011) By IEEE Societies, February 28- March 03, 2011,Chennai, Tamil Nadu, India. ISBN 978-87-92329-61-5, 2011

[9]   Mandal J. K., Sarkar Arindam, "Neural Network Guided Secret Key based Encryption through Cascading Chaining of Recursive Positional Substitution of Prime Non-Prime (NNSKECC)", International Confference on Computing and Systems, ICCS – 2010, 19–20 November, 2010,Department of Computer Science, The University of Burdwan, Golapbag North, Burdwan – 713104, West Bengal, India.ISBN 93-80813-01-5, 2010

[10]   R. Mislovaty, Y. Perchenok, I. Kanter, and W. Kinzel. Secure key-exchange protocol with an absence of injective functions. Phys. Rev. E, 66:066102,2002.

[11]   A. Ruttor, W. Kinzel, R. Naeh, and I. Kanter. Genetic attack on neural cryptography. Phys. Rev. E, 73(3):036121, 2006.

[12]   A. Engel and C. Van den Broeck. Statistical Mechanics of Learning. Cambridge University Press, Cambridge, 2001.

[13]   T. Godhavari, N. R. Alainelu and R. Soundararajan "Cryptography Using Neural Network " IEEE Indicon 2005 Conference, Chennai, India, 11-13 Dec. 2005.

[14]   Wolfgang Kinzel and Ido Kanter, "Interacting neural networks and cryptography", Advances in Solid State Physics, Ed. by B. Kramer   (Springer, Berlin. 2002), Vol. 42, p. 383 arXiv- cond-mat/0203011, 2002

[15]   Wolfgang Kinzel and Ido Kanter, "Neural cryptography" proceedings of the 9th international conference on Neural   Information processing(ICONIP 02).

[16]   Dong   Hu   "A   new   service   based   computing   security   model   with   neural cryptography"IEEE07/2009.

17.   Sarkar, A, Mandal , J. K. , " Secured wireless communication through Simulated Annealing Guided Triangularized Encryption by Multilayer Perceptron generated Session Key(SATMLP)", CCSIT-2013, Proceedings of Third International Conference on Computer Science & Information Technology(CCSIT-2013), Computer Science & Information Technology (CS & IT) ISSN : 2231 - 5403 [Online].

8

Design of Star-Shaped Microstrip Patch Antenna for Ultra Wideband (UWB) Applications

Mustafa Abu Nasr[1], Mohamed K. Ouda[2] and Samer O. Ouda[3]

[1] Engineering Department , Al Azhar University, Gaza, Palestine,
`mustafa.abunasr@gmail.com`

[2] Electrical Engineering Department, Islamic University of Gaza, Gaza, Palestine,
`mouda@iugaza.edu.ps`

[3] Electrical Engineering Department, Islamic University of Gaza, Gaza, Palestine
`s.ouda00@hotmail.com`

ABSTRACT

The design and analysis of a new ultra wideband microstrip antenna for optimum performance that satisfied a large bandwidth starting from 3.9GHz to 22.5GHz is introduced . The UWB antenna is capable of operating over an UWB as allocated by the Federal Communications Commission (FCC) with good radiation properties over the entire frequency range. The techniques of enhancing the bandwidth of microstrip UWB antenna were utilized to enhance the performance of the designed antenna. The effect of shifting feed line from the center of patch to the edges was studied in addition to the effect of changing the length of the ground plane. The antenna was designed and simulated using High Frequency Structure Simulator HFSS software packages..

KEYWORDS

UWB, Microstrip line feed, Patch antenna, offset feed.

1. INTRODUCTION

The Federal Communication Commission (FCC) specified some rules for Ultra Wideband (UWB) antenna implementations. It specified the antenna impedance bandwidth form 3.1GHz to 10.6 GHz and any signal that occupies at least 500MHz spectrum can be used in UWB systems[1]. UWB technology can be considered as the most promising wireless technologies that guarantee to provide high data rate transmissions, low complexity, very low interference and easy connection in many different devices such a laptop, digital camera, and high definition TV. Furthermore, it allows the industry to provide a greater quality of services to the end users. High-performance printed circuit board antennas are essential in portable systems [2].The microstrip antennas are considered to be a key component for these applications due to its advantages such as, low profile, low cost, ease of integration with microwave integrated circuits (MIC) and light weight. It consists of a perfect conducting patch over a thin dielectric material called the substrate that is placed above a ground plane. There are many different patch shapes such as the rectangular, circular, elliptic, circular ring, triangular and hexagonal. There are various techniques for feeding the antenna such as microstrip line feed, coaxial probed feed, aperture-coupled and electromagnetically coupled [3]. The major disadvantages of a microstrip antennas are low power handling capability and narrow bandwidth[4]. There are continuous works for increasing the bandwidth using different techniques. One of them a technique with less complexity in structure depends on small cut on the outskirts at the upper and lower edge of patch [5]. The used of two steps at the lower edge of patch, a single horizontal rectangular slot on the patch also give good results [6]. Recently other techniques have been examined to enhance the UWB antenna using of modified shape of short ground plan. Also making multiple rectangular slots at top side of the ground plane enhances the bandwidth of antenna [7]. A larger patch with an etched slot at the lower edge of antenna with vertical slot on patch and small cut on ground plane so modified the impedance bandwidth of antenna [8]. All these techniques and others are based on the modification of the surface current destruction to enhance the antenna bandwidth.

In this paper, a microstrip star-shape antenna was studied using Ansoft's HFSS software package. HFSS is a full-wave electromagnetic simulator based on the finite element method that is considered to be the industry standard for electromagnetic and antenna simulations [9]. The antenna consists of a star-shape radiating element with a partial ground plane and a microstrip line feed from the edge of the patch. The parameters structure of the antenna was optimized to achieve the widest antenna bandwidth and impedance matching.

2. STAR-SHAPED MICROSTRIP PATCH ANTENNA

2.1 Antenna Structure

The structure and dimensions of the proposed antenna are given in Figure 1. A 50Ω microstrip feed line is printed on the top of "Arlon DiClad 880 (tm)" substrate .The substrate has a thickness of h=1.9mm and a relative permittivity ε_r = 2.2, l_s=25mm and w_s=15mm denoting the length and the width of the substrate, respectively. The width of the microstrip feed line is fixed at w_f =4.7mm to achieve 50Ω impedance. On the other side of the substrate, the conducting ground plane has a length of l_g=7.1mm and width w_g the same of substrate width.

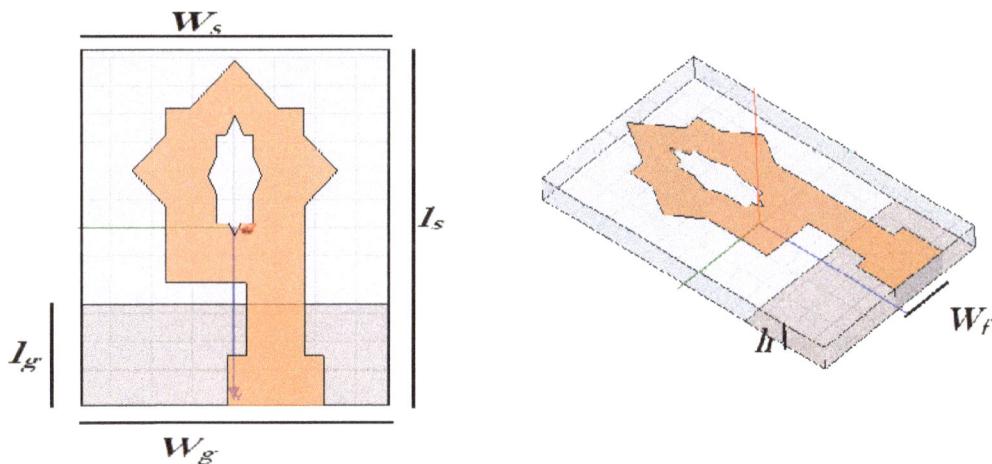

Figure 1. star shape antenna

2.2 Result and Discussion
2.2.1 Antenna Bandwidth

Frequency bandwidth (BW) is the range of frequencies within which the performance of the antenna, with respect to some characteristic, conforms to a specified standard. The frequency bandwidth of an antenna can be expressed as either absolute bandwidth (ABW) or fractional bandwidth (FBW) as shown in equation 1 and 2

$$ABW = f_H - f_L \tag{1}$$

$$FBW = 2\left(\frac{f_H - f_L}{f_H + f_L}\right) \tag{2}$$

Where f_H, f_L and f_c denote the upper edge ,lower edge and center frequency respectively. For broadband antennas, the bandwidth can also be expressed as the ratio of the upper to the lower frequencies, where the antenna performance is acceptable, as shown in equation 3[1].

$$BW = \left(\frac{f_H}{f_L}\right) \tag{3}$$

The bandwidth of antenna can be considered as the range of frequencies bounded by the S_{11} that is lower than -10dB and it can be calculated from the S_{11}(dB) plot [1]. Figure 2 shows that the bandwidth covering an extremely wide frequency range from 3.9GHz to 22.5GHz with a operating frequency of 10.29 GHz.

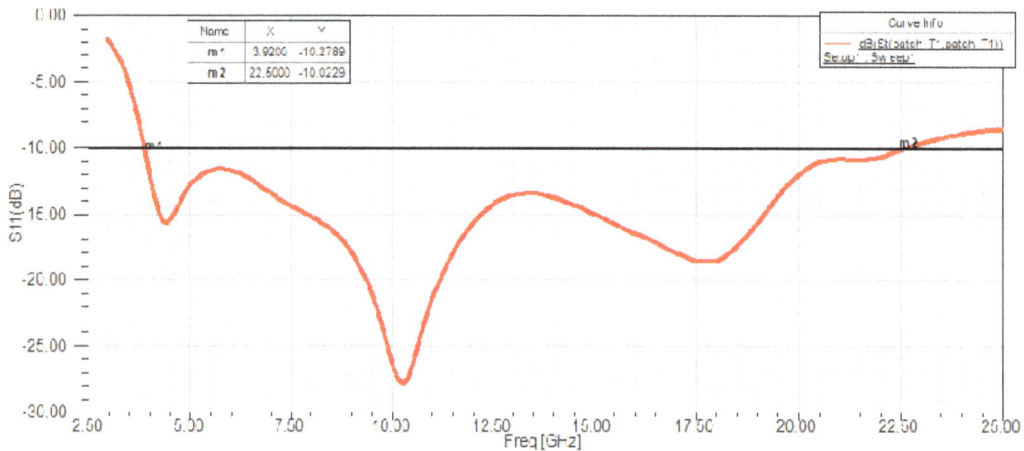

Figure 2. S_{11} (dB) of antenna

2.2.2.Antenna Impedances

The microstrip transmission line feed was selected as feeding techniques for desired antenna. In this method a conducting strip connects directly to the edge of microstrip patch. The advantage of this technique comes from that layout, where the feed can be etched to same substrate to provide a planar structure [10]. The most challenge in design microstrip line is to calculate the width of line that provides good impedance matching. There is a relatively straight forward equation to calculate the characteristic impedance Z_0 given, feed line width w_f, substrate height h, and effective dielectric constant ε_{eff} [3].

For $\dfrac{w}{h} \geq 1$

$$\varepsilon_{eff} = \frac{(\varepsilon_r + 1)}{2} + \frac{(\varepsilon_r - 1)}{2}\left[1 + 12\frac{h}{w_f}\right]^{-1/2} \qquad (4)$$

$$Z_0 = \frac{120\pi}{\sqrt{\varepsilon_{eff}}\left[\dfrac{w_f}{h} + 1.393 + 0.667\ln\left(\dfrac{w_f}{h} + 1.444\right)\right]} \qquad (5)$$

From equation 4 and 5 we found that to achieve $50\,\Omega$ characteristic impedance the w_f is equal to 5.8mm, but the optimized value using HFSS software we found that w_f =4.7mm.

Figure 3 shows the resistance and reactance behavior of the antenna as a function of frequency. It shows that the low S_{11} (<-10dB) always occurs over the frequency range when the input impedance is matched to $50\,\Omega$, i.e. the input resistance R is close to $50\,\Omega$ while the input reactance X is not far from zero. At operating frequency, the resistance is closed to $50\,\Omega$ and the reactance closed to 0, thus resulting in an impedance matching at the antenna and hence increasing of the operating bandwidth.

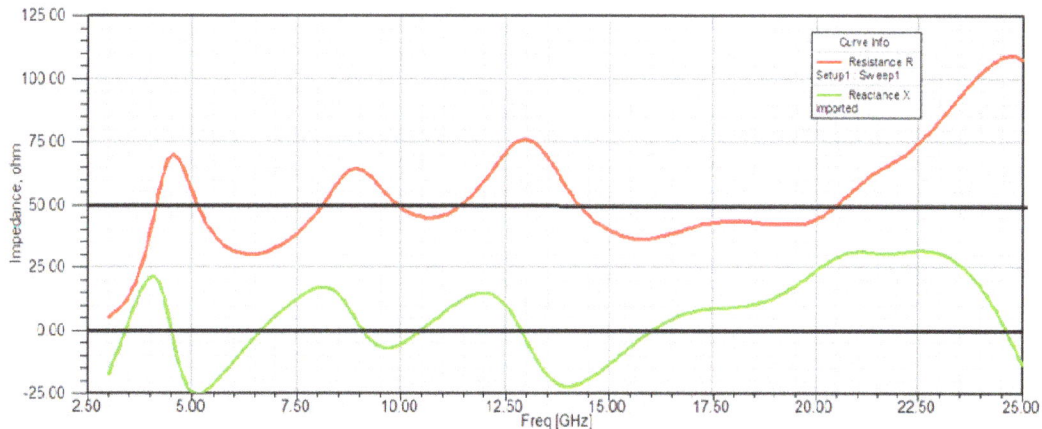

Figure 3. Resistance R and Reactance X

2.2.3 Antenna Gain

The simulated of maximum gain of the optimized antenna at $\phi = 0^o, 90^o$ and $\theta = 0° - 180°$ as a function of frequency is illustrated in Figure(4). It shows that at 10.29 GHz and 17.8GHz frequency the gain is 4.48dB and 4.4dB respectively at $\phi = 90^o$, but at $\phi = 0^o$ the gain is 1.95dB and 5dB for both 10.29 GHz and 17.8GHz frequency respectively.

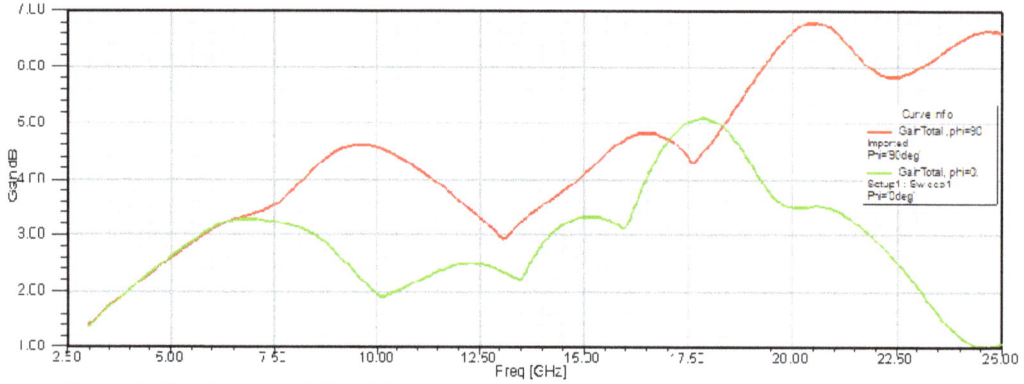

Figure 4. The simulated Gain of the antenna at $\phi = 0°, 90°$ and $\theta = 0° - 180°$ over frequency band

2.2.4   Antenna Radiation pattern

Figure 5 illustrates the simulated of radiation pattern for both frequency 10.29GHz and 17.8GHz at $\phi = 0°, 90°$ and, $\theta = 0° - 180°$

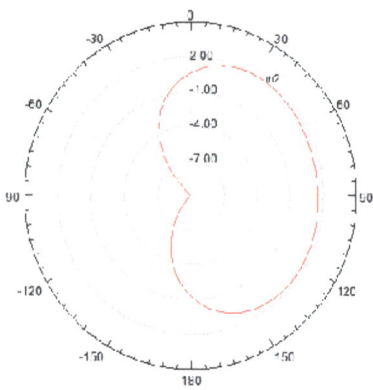

Figure5.a. radiation pattern at 10.2GHz, $\phi = 0°$

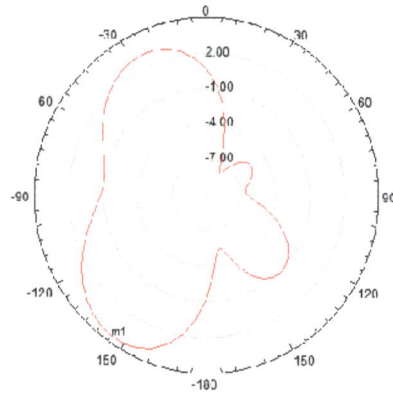

Figure5.b. radiation pattern at 10.2GHz, $\phi = 90°$

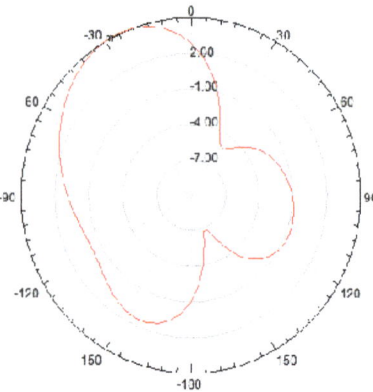

Figure 5.c. radiation pattern at 17.8GHz, $\phi = 0°$

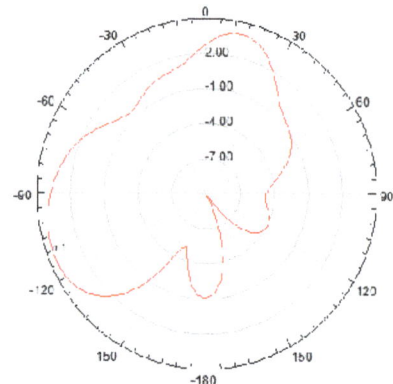

Figure 5.d. radiation pattern at 17.8GHz, $\phi = 90°$

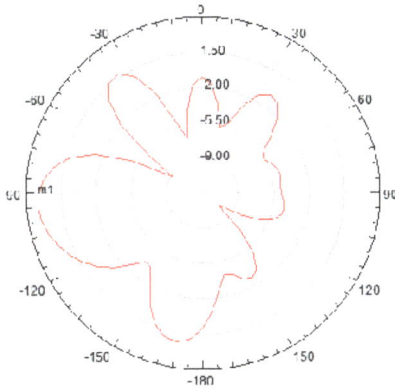

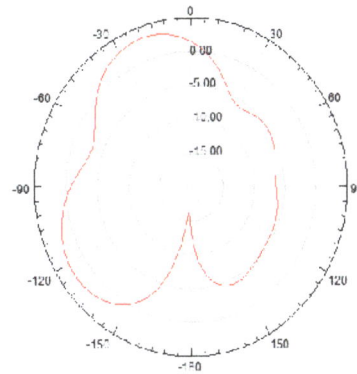

Figure5.e. radiation pattern at 17.8GHz, $\theta = 0^o$ Figure 5.f. radiation pattern at 10.19GHz, $\theta = 90^o$

From Figure 5.a-b we can notice that the radiation pattern at 10.9GHz when $\phi = 90^o$ is more directional compare to that at $\phi = 0^o$. Also from Figure (5.c-d) it is shown that the radiation pattern at 17.8GHz, when $\phi = 0^o$ is more directional than that at $\phi = 90^o$ Figure 5.e-f show the horizontal plane of radiation pattern at $\theta = 90^o$.

2.3 Parametric Study

A parameter study was conducted to optimize antenna parameters. It helps to investigate the effect of different parameter on the impedance bandwidth. The effect of feed line shift of microstrip line and the ground plane length are studied. All the antenna parameters were kept constant in the simulation except for the parameter of interest.

2.3.1 Effect of feed shift(Offset feed)

Figure6 illustrates the simulated of S11(dB) for different feed shift steps for microstrip line feeder from the center of radiating element to its edge when the ground plan width equal to w_g =15 mm and its length is fixed at l_g =7.1mm. Shift steps of   =0, 0.4, 0.8, 1.2, 1.6, and 2mm were simulated.

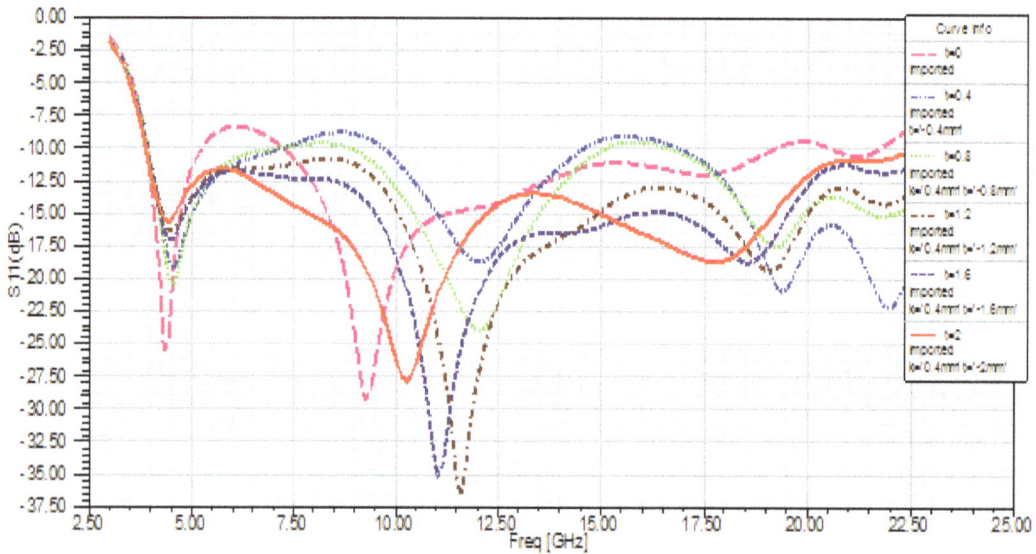

Figure 6. Simulated of S_{11}(dB) curves for different feed shift

It is shown in Figure 6 that the -10dB operating bandwidth of the antenna varies remarkably with the variation of the feed shift t. The optimal feed shift is found to be 2mm with the bandwidth covering an extremely wide frequency range from 3.9GHz to 22.5GHz. The center frequency 10.29 GHz is obtained which is very close to the desired frequency of operation. It was observed from many trials of simulations that as the feed line location is moved away from the center of the patch, the center frequency starts to decrease and the bandwidth increase.

The current distributions on the patch for each feed shift are shown in Figure 7.

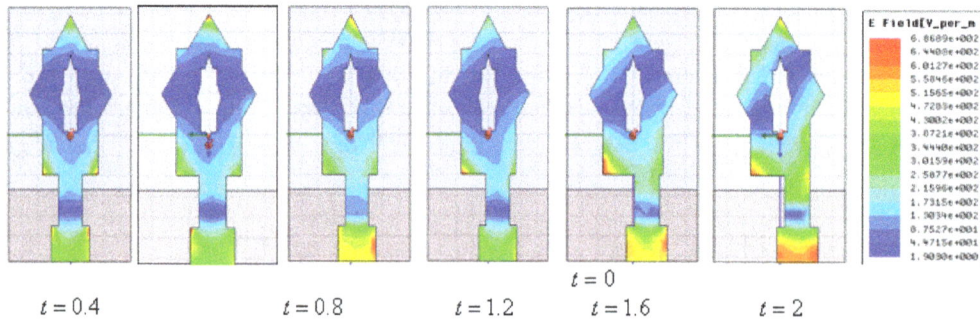

Figure 7. feed shift steps for microstrip-fed line with current distribution on patch

2.3.2 Effect of the Ground Plane

In previous section, it has been demonstrated that the variation of the feed shift t leads to the variations of the frequency bandwidth. In a broad sense, the ground plane serves as an impedance matching circuit and also it tunes the resonant frequencies [1]. To conform this, Figure 8 is given were the simulated S11(dB) curve for the antenna with different ground lengths ($l_g = \lambda$, 0.75λ, $0.5\lambda\,0.25\lambda$).

Figure 8. Simulated of S_{11}(dB) curve for different ground plan length

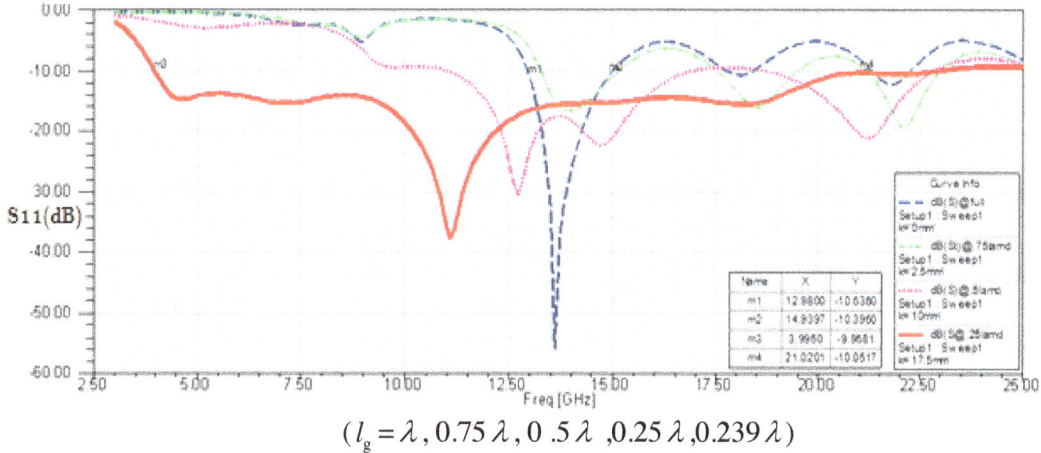

$(l_g = \lambda , 0.75\lambda , 0.5\lambda , 0.25\lambda , 0.239\lambda)$

It is noticed in Figure 8 that for full ground plan the first-10dB bandwidth ranges from 12.9GHz to 14.9GHz, which is much narrower than that of a short ground plane. This is due to the impedance mismatch over an extremely frequency range resulting from the full ground plane. When the length of ground plane starts to decrease, l_g =0.75λ , 0.5λ we see that the bandwidth range starts to increase gradually and the S11 under -10dB curve becomes wider at l_g =0.25λbandwidth start from 3.9GHz to 21.GHz. The optimum values was found by optimizing the antenna is l_g =0.239λ−7.1mm were the bandwidth ranges from 3.9GHz to 22.5GHz with little frequency shift. Table 1 shows the value of the bandwidth at different length of the ground plane

l_g	Start frequency GHz	End frequency GHz	Bandwidth GHz
λ	12.9	14.9	2
0.75λ	13.3	14.9	1.6
0.5λ	11	17.5	6.5
0.25λ	3.9	21	17.1
0.239λ	3.9	22.5	18.6

Table 1. The value of bandwidth at different length of ground plane

3.CONCLUSION

In this paper, a design of new microstrip UWB antenna with good performance was proposed. The antenna was designed and simulated using Ansoft's HFSS electromagnetic simulation package. The effects of feeding shift technique and the truncation of the ground plan method were studied. Enhance the antenna performance parameters was achieved. An extreme antenna bandwidth of 18.6GHz was achieved using ground plan 7.1mm.

REFERENCES

[1]   Jianxin Liang(2006) Antenna Study and Design for Ultra Wideband Communication Applications, a thesis submitted of Doctor of Philosophy, University of London, United Kingdom,

[2]   XiulongBao, Max Ammann: Printed band-rejection UWB antenna with H-shaped slot,Dublin Institute of Technology, 2007-01-01.

[3]   Constantine A. Balanis(2005) Antenna theory analysis and design, Wiley & Sons, Inc , third edition, Canada.

[4]   David R. Jackson: Antenna Engineering Handbook, University of Houston

[5]   M. Y. Alhefnawy, Aladdin Assisi, HosnyAlmotaafy, A. Safwat and M.I. Youssef (2009) "Design and Implementation of a Novel Planer UWB Monopole Antenna for Multipath Environments" ,13th International Conference on Aerospace Sciences& Aviation Technology, ASAT.

[6]   Seok H. Choi, Jong K. Park, Sun K. Kim, and Jae Y. Park, 4 August 2003: A new ultra wide band antenna for UWB application, Department of Radio Wave, Engineering Hanbat National University.

[7]   Mohammed tariqual ,rezaualazim ,norbahian misran ,kamarulzaman mat ,and badrainbais(2010)" Design and optimization printed rectangular antenna for UWB application", World applied sciences .

[8]   Lee Chia Ping, Chandan Kumar Chakrabarty andRozanah Amir Khan,(2010) "Enhanced Bandwidth of Impulse-Ultra Wideband (I-UWB) Slotted Rectangular Patch Antenna with Partial Ground Plane" IJECCT.

[9]   Ansoft Corporation, user's guide: High Frequency Structure Simulator V10.

[10]   Yogesh K. Choukiker, (2009)" Analysis of dual band rectangular microstrip antenna using IE3D/PSO", Master of technology in Telematics and signal processing, National Institute of Technology, 2009

9

DATA COLLECTION SCHEME FOR WIRELESS SENSOR NETWORK WITH MOBILE COLLECTOR

Khaled Almi'ani, Muder Almi'ani, Ali Al_ghonmein and Khaldun Al-Moghrabi

Al-Hussein Bin Talal University, Ma'an, Jordan

ABSTRACT

In this paper, we investigate the problem of designing the minimum number of required mobile elements tours such that each sensor node is either on the tour or one hop away from the tour, and the length of the tour to be bounded by pre-determined value L. To address this problem, we propose heuristic-based solution. This solution works by directing the mobile element tour towards the highly dense area in the network. The experiment results show that our scheme outperform the benchmark scheme by 10% in most scenarios.

1. INTRODUCTION

Typically, in multi-hop data communication scenario, the sensors closed to the sink are the first to run out energy. This is due to fact that these sensors are required to forward all other nodes data. Once these nodes are disconnected, the operational lifetime of the networks ends, since the entire network become unable to communicate with the sink. This raise reducing the energy consumption as a major challenge.

To address this problem, many proposals [1][2][3] in the literature, have investigated the use of the Mobile Elements (MEs). A Mobile Element serves as a data carrier and travel the entire network to collect the data of each sensor via single-hop communication. By using Mobile Elements, we can avoid the energy consumption due to multi-hop communication. However, the speed of the mobile element is typically low[4][5], and therefore reducing the data gathering latency is important to ensure the efficiency of such approach.

In this paper, we assume that are given more than one mobile element and the objective is to design minimum number of required tours to cover the network. Such that each sensor node is either on the tour or one hop away from the tour, and the length of the tour to be bounded by pre-determined value *L*. We term this problem as the Length-constrained Tours Minimization. To address the presented problems we present heuristic-based solution, which uses cluster-based mechanism to build it solutions.

The rest of the paper is organized as follows. Section 2 presents the related work in this area. Section 3 provides formal definitions of the problem presented in this work. Section 4 presents the heuristic solution of the presented problem. In Section 5 the evaluation for the proposed heuristics is presented. Finally, Section 6 concludes the paper.

2. RELATED WORK

In the literature, several proposals[6][7] have investigated the use mobile sink(or mobile elements) to increase the lifetime of the network. By varying the data gathering path, , the residual energy in the nodes becomes more evenly balanced throughout the network, leading to a higher network lifetime.

Zhao et al. [8][9] investigated the problem of maximizing the overall network utility. Accordingly, they proposed two algorithmic-based solutions, where the mobile element visit each gathering point for a period of time and gather the data from nearby sensors via multi-hop communications. They considered the cases where the sojourn time is fixed as well as variable. Guney et al. [10] formulated finding the optimal sink trajectory and the data flow routs problem as mixed integer programming formulations. Accordingly, they presented several heuristic to address the presented problem.

The problem presented in this work share some similarities with the problem proposed by Xu et al.[11]. In their problem, the objective is to plan the mobile element tour to visit set of nodes in the network named gathering points. The upper-bound of the obtained tour must be also bounded by pre-determined time-deadline. In addition, they also restrict the depth of the routing trees.

The problems presented in this paper inherits some characteristic from the well-known Vehicle Routing Problem (VRP) [12]. In this problem, given a group of vehicles assigned to an initial point, the goal is to determine routes for these vehicles from the initial point to the customers to deliver goods while minimizing the travelling time for these vehicles. Also, our problem can be recognized as variation of the Deadline Travelling Salesman Problem (Deadline-TSP)[13]. In this problem the objective is determine the shortest tour for a salesman to visit a set of cities, where each city must be visited before a time deadline.

The single hope data gathering problem proposed by Ma et al.[14] can be recognized as an extended version of the problems presented in this work. In[14] , the authors investigated the problem of designing the mobile element tour(s), such that each node is either on the tour or one hop away from the tour. The only difference between this problem and the problems presented in this is the introduction of the "polling points" by the authors. Ma et al.[14] defines the polling point as a point in the network, where the mobile element can communicate with one or more sensor node via a single hop transmission. In situations, where the network is spars, where the polling points are the sensor nodes, the problems presented in this work become exactly similar to the problem presented by Ma et. al.[14].

3.Problems Definitions

In the Length-constrained Tours Minimization problem, we are given an undirected complete graph $G = \langle V, E \rangle$. V is the set that represents the sensor nodes in the network ($v_s \in V$, v_s in the sink node) . E is the set of edges that represents the travelling time between the node in the network, i.e. (v_i, v_j) is the time that the mobile element takes to travel between nodes v_i and v_j. Also we are given M and L. M is the set that represents the available mobile elements, and L is the time-deadline constraint.

A solution for this problem consists of multiple tours, where the objective is to minimize the number of obtained tours, such that the travelling time for each tour is less than or equal to L, and each node is on one of the tours or one-hop away from a node included in one of the tours.

4.Algorithmic solution

To address the presented problem, we present heuristic approach that works by partitioning the network, then in each partition, a mobile element will be assigned. This partitioning takes into consideration the distribution of the nodes, to avoid long distance travelling by the mobile element. The presented approach uses ideas from [15][16][17].

Our approach start by identifying the set of nodes that will be used to construct the mobile elements tours. Nodes does not belong to this set must be at most one hope away from nodes belong to this set. To obtain this set we use the Set-Coverd-based algorithm proposed by Almi'ani et al[15]. Once the nodes of the tours are identified, the presented approach start by partitioning the network into two partitions. The partitioning step employs the well-known k-mean algorithm. Once these two partitioned are obtained, the process proceed by constructing a single tour for each partition. The tours inside each partition is constructed using Christofides algorithm[18]. For each partition, if the obtained tour satisfies the time-deadline constraint, this tour will be assign to a mobile element. Otherwise, this partition will be re-partitioned and the tour construction step will be retriggered.

The k-mean algorithm aims to partition the network graph into k number of clusters such that the distance between the nodes inside each clusters is minimized. At first, k number of nodes are selected at random as the initial center nodes of the clusters. Then, in each iteration, each node is assigned to its nearest cluster (center node). Once all nodes have been thus assigned, the center node for each cluster is recalculated, and the process is repeated from the beginning based on the identity of the new center nodes. The clustering step stops when the obtained centre nodes still the same in two consecutive iterations.[16]

5. EXPERIMENT RESULTS

To validate the performance of our algorithm, we have conducted an extensive set of experiments. We aim to investigate the performance of the presented heuristic in terms of the number of obtained tours and length of the tours. We consider varying the following parameters in our experiments:

(1) varying the number of nodes in the network and
(2) varying the value of the tour length constraint.

 Each of the presented experiment is an average of 10 different random topologies. We consider the following deployment scenarios:

- Uniform density deployment: the nodes are uniformly deployed in a square area of $400 \times 400 \ m^2$.
- Variable density deployment: the area of the network is divided into 25 square, where the size of each square is $80 \times 80 \ m^2$. Then, ten of these squares are selected randomly, and the density of the selected squares are set be 5 times the density in the remaining squares.

In this section we will refer to our algorithm as the Partition-Based algorithm. To benchmark the presented algorithms performances, we compare the Partition-Based algorithm against the Data Gathering Algorithm with Multiple M-collector, which also proposed by Ma et al.[14], we will refer to this algorithm as the M-Covering algorithm.

In the M-Covering algorithm, given the polling points and the transit constraint, this algorithm starts by constructing a minimum spanning tree that connects all polling points. Then, for each node in the tree we calculate the total calculate the summation of the lengths of the edges in the sub-tree rooted at this node. Then, in each iteration, we extract the deepest, heaviest sub-tree, where the total edges weight of this sub-tree is less than or equal to L/2. Once a sub-tree is extracted, the weight value for each node left in the original tree is re-calculated. This process stops when the tree become empty. The tour for each extracted sub-tree is calculated by using a TSP-solver. The bound is used because the authors employed a 2-approximation algorithm [19] to obtain the tour. In this section, to ensure the fairness of the evaluation, we employed Christofides algorithm as a TSP-solver in the M-Covering algorithm. Also we replace the condition <=L/2, by continuously validate that the length of the obtained tour is L[17].

In this section, we compare the Partition-Based (PB) algorithm against the M-Covering algorithm. We are particularly interested in investigating the performance of the algorithms, while varying the number of nodes and the time deadline constraint (L). Unless mentioned otherwise the value of $L = d \times 1 \, m/s$, where d is the distance between the sink and the farthest node in the network. Now, we move to investigate the relationship between the number of nodes and the number of tours obtained by each algorithms. Figures 7 and 8 show the results for the uniform density and the variable density deployment scenarios; respectively. From the figures, we can see that the Partition-Based algorithm always obtains lower number of tours, compared to the M-Covering algorithm. Also, from the figures we can see that in the variable density deployment scenario, the M-Covering algorithm obtains more tours compared to the uniform deployment scenario. The factors behind this behavior are main the process of selecting the tours' nodes, and the partitioning step. The M-Covering algorithm uses the process employed by the T-Covering algorithm to identify the nodes involved in the mobile elements tours. In addition, the Partition-Based algorithm uses the same process used by the Set-Based algorithm to identify the nodes included in the mobile elements tours. In this direction, we expected that number of nodes selected to be included in the tours to be significantly bigger in the M-Covering algorithm, compared to the Partition-based algorithm. The partitioning process employed by the M-Covering algorithm works by cutting the minimum spanning tree, which connect the identified nodes, from bottom to top into smaller trees that satisfy the time deadline constraint. Such cutting mechanism works without considering the distance between the nodes, which are packed into one tour. This is expected to reduce performance of such a mechanism; especially when the distances between neighboring nodes vary frequently, since in this situation, traversing the minimum spanning tree may result in having a few nodes in each tour, which very far from each other. In the Partition-based algorithm, the obtained tours' nodes are partitioned using the k-mean algorithm, and this algorithm has the advantage of partitioning the nodes into clusters (groups of nodes) such that the distances among the same cluster nodes are relatively small. This gives the Partition-based algorithm the advantage, since any tour's nodes are not expected to be far from each other.

Now, we study the impact of the time deadline constraint (L) on the number of tours obtained by each algorithm. Figures 7 and 8 show the results for both deployment scenarios, for 500-nodes networks. Here, the horizontal axis shows the value of L normalized as a fraction of $d \times 1 \, m/s$. We observe that, reducing the value of L reduces the gap between the presented algorithms performances. This is expected, since reducing the value of L reduces the solution space and therefore reduces the effect of each algorithm main advantages.

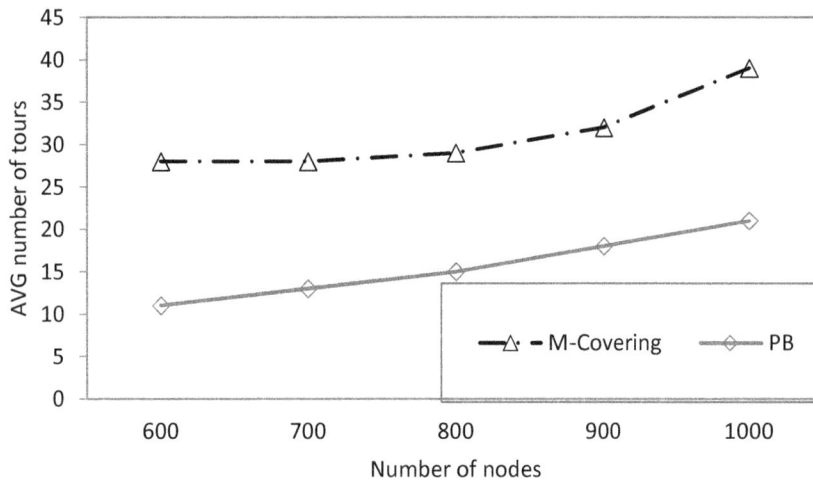

Figure 5: Number of nodes against total the average number of tours, for the uniform density deployment scenario.

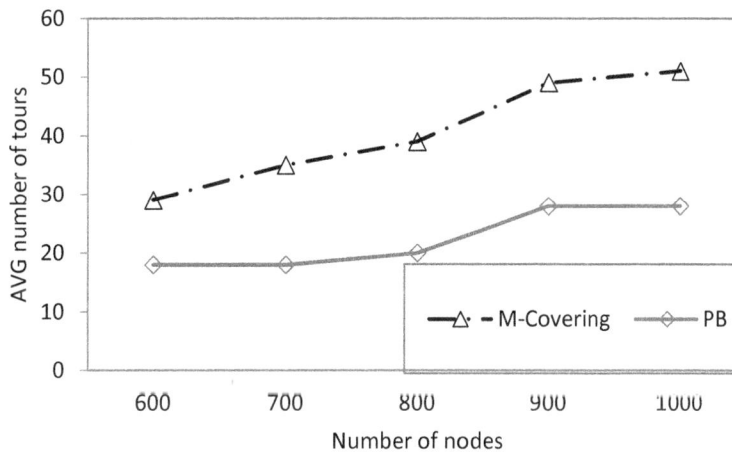

Figure 5: Number of nodes against total the average number of tours, for the variable density deployment scenario.

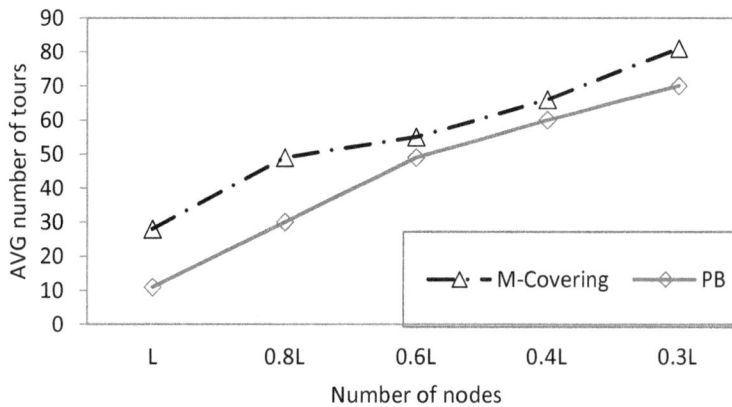

Figure 5: Normalized value of L against the average number of tours, for the uniform density deployment scenario.

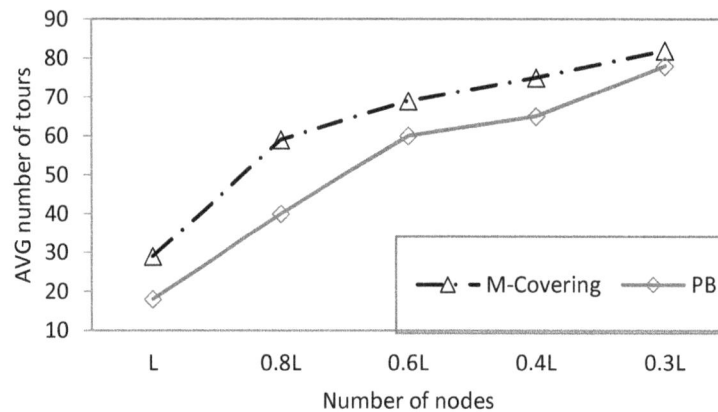

Figure 5: Normalized value of L against the average number of tours, for the variable density deployment scenario.

6.CONCLUSION

In this paper, we consider the problem of designing the mobile elements tours such that total size of the routing trees is minimized. In this work, we present an algorithmic solution that creates its solution by partitioning the network, then in each partition; a caching node is selected based on the distribution of the nodes. The algorithms presented here are based only on the static network topology. All nodes are considered to be identical in terms of energy consumption and communication capabilities. In order to further maximize the network lifetime, our algorithms could be extended to take into account the residual energy of individual nodes, during the design of the tours.

REFERENCES

[1]   K. Almi'ani, S. Selvadurai, and A. Viglas, "Periodic Mobile Multi-Gateway Scheduling," in Proceedings of the Ninth International Conference on Parallel and Distributed Computing, Applications and Technologies (PDCAT), 2008, pp. 195–202.

[2]   Y. Gu, D. Bozdag, E. Ekici, F. Ozguner, and C. G. Lee, "Partitioning based mobile element scheduling in wireless sensor networks," in Sensor and Ad Hoc Communications and Networks, 2005. IEEE SECON 2005. 2005 Second Annual IEEE Communications Society Conference on, 2005, pp. 386–395.

[3]   G. Xing, T. Wang, Z. Xie, and W. Jia, "Rendezvous planning in wireless sensor networks with mobile elements," IEEE Trans. Mob. Comput., vol. 7, no. 12, pp. 1430–1443, Dec. 2008.

[4]   K. Dantu, M. Rahimi, H. Shah, S. Babel, A. Dhariwal, and G. S. Sukhatme, "Robomote: enabling mobility in sensor networks," in Proceedings of the 4th international symposium on Information processing in sensor networks (IPSN), 2005, pp. 404 – 409.

[5]   R. Pon, M. A. Batalin, J. Gordon, A. Kansal, D. Liu, M. Rahimi, L. Shirachi, Y. Yu, M. Hansen, W. J. Kaiser, and others, "Networked infomechanical systems: a mobile embedded networked sensor platform," in Proceedings of the 4th international symposium on Information processing in sensor networks, 2005, pp. 376 – 381.

[6]   S. R. Gandham, M. Dawande, R. Prakash, and S. Venkatesan, "Energy efficient schemes for wireless sensor networks with multiple mobile base stations," in Proceedings of IEEE Globecom, 2003, vol. 1, pp. 377–381.

[7]   Z. M. Wang, S. Basagni, E. Melachrinoudis, and C. Petrioli, "Exploiting sink mobility for maximizing sensor networks lifetime," in Proceedings ofthe 38th Annual Hawaii International Conference on System Sciences (HICSS), 2005, vol. 9, pp. 03–06.

[8]   M. Zhao and Y. Yang, "Optimization-Based Distributed Algorithms for Mobile Data Gathering in Wireless Sensor Networks," IEEE Trans. Mob. Comput., vol. 11, no. 10, pp. 1464–1477, 2012.

[9]   M. Zhao and Y. Yang, "efficient data gathering with mobile collectors and space-division multiple access technique in wireless sensor networks," IEEE Trans. Comput., vol. 60, no. 3, pp. 400–417, 2011.

[10] E. Güney, I. K. Altmel, N. Aras, and C. Ersoy, "Efficient integer programming formulations for optimum sink location and routing in wireless sensor networks," in Proceedings of the 23rd International Symposium on Computer and Information Sciences, 2008, pp. 1–6.

[11] Z. Xu, W. Liang, and Y. Xu, "Network Lifetime Maximization in Delay-Tolerant Sensor Networks With a Mobile Sink," in Proceedings of the 8th IEEE International Conference on Distributed Computing in Sensor Systems, 2012, pp. 9–16.

[12] P. Toth and D. Vigo, "The Vehicle Routing Problem," Soc. Ind. Appl. Math., 2001.

[13] N. Bansal, A. Blum, S. Chawla, and A. Meyerson, "Approximation algorithms for deadline-TSP and vehicle routing with time-windows," in Proceedings of the thirty-sixth annual ACM symposium on Theory of computing (STOC), 2004, pp. 166 – 174.

[14] M. Ma, Y. Yang, and M. Zhao, "Tour Planning for Mobile Data-Gathering Mechanisms in Wireless Sensor Networks," IEEE Trans. Veh. Technol., 2013.

[15] K. Almi'ani and A. Viglas, "Designing connected tours that almost cover a network," in Proceedings of the 2013 14th International Conference onParallel and Distributed Computing, Applications and Technologies(PDCAT'13), 2013, pp. 281 – 286.

[16] K. Almi'ani, A. Viglas, and L. Libman, "Energy-efficient data gathering with tour length-constrained mobile elements in wireless sensor networks," in Proceeding of the IEEE 35th Conference on Local Computer Networks (LCN), 2010, pp. 582–589.

[17] K. Almi'ani, M. Aalsalem, and R. Al-Hashemi, "Data gathering for periodic sensor applications," in in Proceedings of the 12th International Conference on Parallel and Distributed Computing, Applications and Technologies, 2011, pp. 215–220.

[18] N. Christofides, "Worst-case analysis of a new heuristic for the traveling salesman problem," 1976.

[19] S. S. Skiena, "Traveling Salesman Problem," Algorithm Des. Man., pp. 319–322, 1997.

10

VIRTUAL 2D POSITIONING SYSTEM BY USING WIRELESS SENSORS IN INDOOR ENVIRONMENT

Hakan Koyuncu[1] and Shuang Hua Yang[2]

[1] Computer Science Department, Loughborough University, Loughborough, UK
[2] Computer Science Department, Loughborough University, Loughborough ,UK

ABSTRACT

A 2D location detection system is constructed by using Wireless Sensor Nodes (WSN) to create aVirtual Fingerprint map, specifically designed for use in an indoor environment. WSN technologies and programmable ZigBee wireless network protocols are employed. This system is based on radio-location fingerprinting technique. Both Linear taper functions and exponential taper functions are utilized with the received signal strength distributions between the fingerprint nodes to generate virtual fingerprint maps. Thus, areal and virtual combined fingerprint map is generated across the test area. K-nearest neighborhood algorithm has been implemented on virtual fingerprint maps, in conjunction with weight functions used to find the coordinates of the unknown objects. The system Localization accuracies of less than a grid space areproved in calculations.

KEYWORDS

Wireless Sensors Node (WSN), Received Signal Strength (RSS), Link Quality Indicator,(LQI), Application Program (AP), Weight function, Fingerprint, k-NN algorithm, Interpolation function, Virtual Node (VN), ZigBee protocol, Jennic.

1. INTRODUCTION

Wireless Sensor Network technology is used for variety of indoor navigation and position detection, [1]. There are many position identification systems using optical [2,3], ultrasonic [4,5] and RF wireless technologies [6,7]. Each technology has its own characteristics and cost factors. Environmental conditions affect the accuracy of object position detection in different ways with these techniques. RF wireless technology has advantages such as having contactless and none line-of-sight nature and being able to operate in extreme environmental conditions, [8].

In RF based localization which is used in this study, the target object carries a Wireless Sensor acting as a receiver. Radio Signal Strength (RSS) information arrives at the receiver in the form of Link Quality Indicator(LQI) from nearby Wireless Sensors acting as transmitters. This information is transferred to a PC server to calculate the position of the target object. WSN localization uses fingerprint database which is constructed by the measured LQI values. Fingerprint based localization has two operational phases identified as, "off-line" and "on-line" phases. In the off-line phase, the location fingerprint database is organized.

The received LQI values fromthe transmitters via a receiver positioned at each grid point are identified as the location fingerprint of that particular grid point. In the on-line phase, LQI values, arriving from the transmitters, are received by a receiver on the target object at the unknown location. These LQI values are identified as the object fingerprint. Although the Fingerprint model and its database work well with many localization systems, due to its coarse nature and the relatively large distances between its grid points, RF signal receptions at grid points are affected. Each grid point is an RF signal measurement point. Random nature of the RF signal measurements at these points based on enviromental factors alsohas impact on the recordings of the uniform RSS or LQI values and correlation between these measurements are decreased.

A solution is proposed to quantize the signal strengths between the grid points and introduce a fingerprint database with denser measurement values which would reduce the signal strength uncertainties between the grid points. The number of grid points across the test area can be increased in two ways. First approach is to increase the grid points by physicallyincreasing the number of measurement points. This increase results not only in a larger fingerprint database, but in return, takes more time and effort during measurements.

Second approach, on the other hand, keeps the number of grid points the same and introduces virtual grid points between the physical grid points. Hence new LQI measurements are generated virtually and a new, larger fingerprint database is developed with real and virtual LQI measurements across the test area. The new approachdoes not utilize additional transmitters, receivers and grid measurement points. The system deploys only virtual grid points integrated among the physical grid points with a larger number of LQI values across the test area in indoor environments, [9,10].

The rest of this paper is organized as follows: In section 2, a brief overview of fingerprint localization is described. In this section, Euclidian distance calculations and weight functions are introduced between the target object and the virtual grid points. In section 3, Virtual grid generation is described. In this section, virtual LQI calculations are presented with respect to linear and exponential taper functions. In section 4, implementation of this new approach is explained and the results are shown for different fingerprint databases. Finally, in section 5, discussions and conclusions aresummarized.

2. FINGERPRINT LOCALIZATION

In Fingerprintlocalization technique, A number of wirelesssensor transmitters and a receiver on the object are employed across the test area as shown in Figure 1. B_i transmitters where i=1,2,3,4 at known positions transmit their LQI values to receiver. LQI measurements are collectedby a receiverlocatedat every measurement point identified as a grid point. These measurements are recorded in Fingerprint database.

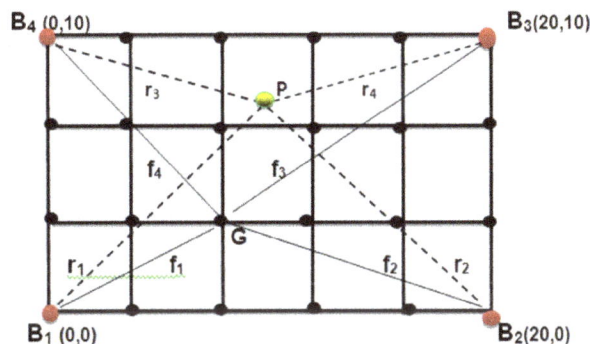

Figure 1: Grid area is showing a grid point G, an unknown point P and B_i transmitters where i=1,2,3,4

The signal distance between the object location P and the grid point G in the fingerprint map is calculated by using LQI values recorded at the respective positions. Fingerprint vector F is identified as the total LQI values from B_i transmitters at a particular grid location G. The vector F is denoted as

F= (f_1, f_2, f_3, f_4)

The unknown object location fingerprint vector R is the LQI values recorded at point P and denoted as

R = (r_1, r_2, r_3, r_4)

Fingerprint database is prepared by collecting F vectors from all the grid points during off-line phase. R fingerprint vector at point P is also recorded and sent to PC during on-line phase, [11,12].The Euclidean distance, E, between F and R vectors at any grid point is given by:

$$E = \left(\sum_{i=1}^{N} (f_i - r_i)^2 \right)^{\frac{1}{2}} \quad (1)$$

whereN is the number of transmitters .k number of smallest Euclidean distances are selected from the total number of Euclidean distances and their corresponding coordinates are averaged out to give the estimated value of the object position coordinates (x,y). This estimation algorithm is called k-nearest neighborhood algorithm (k-NN).To improve the accuracy of the estimation, weight functions are employed with thek-NN algorithm. Object position coordinates (x,y), can be defined by equation 2,

$$(x, y) = \sum_{i=1}^{k} w_i (x_i, y_i) \quad (2)$$

where $x = \sum_{i=1}^{k} w_i . x_i$ and $y = \sum_{i=1}^{k} w_i . y_i$

w_i is the weight function of the ith neighboring grid point in k-nearest neighborhood and (x_i, y_i) is the coordinates of the k-nearest neighborhood grid points. The weight function used in the study, [13], is defined as

$$w_i = \frac{\dfrac{1}{E_i^3}}{\sum_{i=1}^{k} \dfrac{1}{E_i^3}} \quad (3)$$

whereE$_i$is the individual Euclidean distances at k-nearest grid points.

3.VIRTUAL FINGERPRINT

A physical grid system is utilized for indoor localization where LQI measurements are carried out at each grid point and these LQI values with respect to their grid coordinates are defined as physical fingerprint. These grids are organized across the test area with ample distances between them where RF radiation shows characteristics of signal variations. In order to include these effects in localization calculations either several new LQI measurements are taken or a virtual RF signal distribution is assumed between adjacent grid points.

Virtual RF signal strentgh values areassumed at virtual grid points between adjacent grid points following a distribution function. These virtual LQI values with respect to their coordinates are identified as virtual fingerprint.   The location of virtual grid points can be defined emperically between the two adjacent grid points. A physical grid system is organized across the test area as shown in Figure 2. Transmitters are stationed at the corners with known coordinates of the test area. Grid cell size is arranged according to the size of the sensing area.   Radio signals,transmittedfrom B_1,B_2,B_3 and B_4 transmitters, are received by the receiver on the object and at the grid points in the form of LQI values.

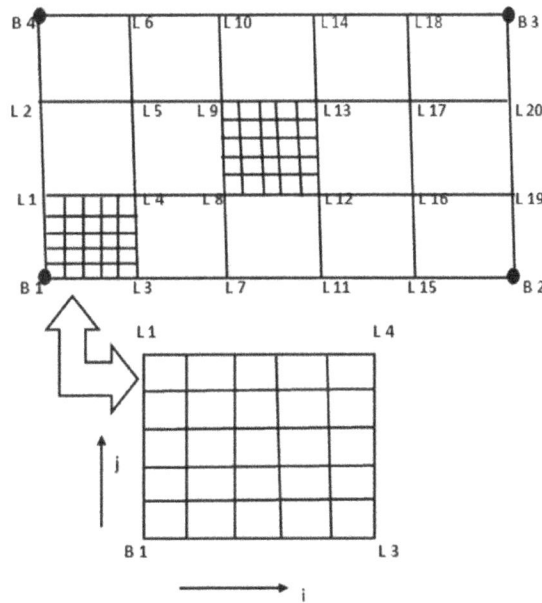

Figure 2.Physical grid space andvirtual grid cells(ie. 5) across each physical grid cell are displayed

Each physical grid cell such as ($L_8L_{12}L_{13}$ L_9) is in square shape and is surrounded by 4 grid points. Each cell is further divided into n^2 square shaped virtual grid cells. n-1 virtual grid points between two adjacent grid pointsaredeployed.The number of physical grid points is decided according to the size and shape of indoors. For representation purposes, the number of virtual grid cells, selected in this study, is n=5.

Since the coordinates of the physical grid points are defined with respect to transmitter positions, the coordinates of the virtual grid points can also be easily calculated. LQI values at adjacent grid points are considered to generate LQI values at virtual grid points between them. The distribution of virtual LQI values between two adjacent grid points are organized according to window functions named as taper functions. Once the LQI measurements arerecorded at the physical grid points, the distribution of virtual LQI values between two adjacent grid points can be assumed according to any distribution function.

Two taper functions are considered: linear taper function and exponential taper function.The distance between two adjacent grid points is divided into n sections. n-1 number of virtual LQI values aregenerated between these two adjacent grid points according to taper functions. Hence the physical grid space is divided into a finer virtual grid space.   Virtual LQI values at virtual grid points and the measured LQI values at grid points produce a new fingerprint database. This fingerprint database is identified as virtual fingerprint database.

Unknown location detection is carried out by using newly generated virtual fingerprint database and k-NN and weightedk-NN algorithms. Virtual euclidean distances areutilized betweenthe object location and the virtual grid points. k number of minimum virtual euclidean distances are selected and their weighted coordinates are averaged out to determine the object location. In conclusion, the positioning technique employed with virtual grid system is same as physical grid system. An overview of the proposed system is presented in a block diagram as shown in Figure 3:

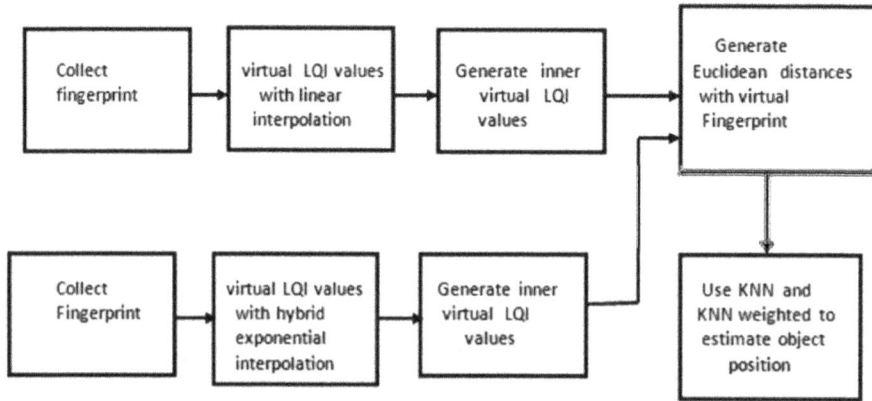

Figure 3. Block diagram of the proposed localization system

It was important that no additional wireless sensor nodes are employed across the test area. The technique introducesan increased number ofLQI values due to denser virtual grid points without any extra effort. Hence, fixed number of grid points and the fixed number of LQI data in the fingerprint database are abandoned in favor of larger number of virtual grid points and virtual LQI data.

3.1. Linear Interpolation Function

The proposed approach uses linear taper function which can be identified as the linear interpolation function of LQI values at virtual grid points between two adjacent physical grid points. Linear taper function is utilized to calculate the virtual LQI values in every virtual grid point.

Each physical grid cell has 4 LQI values at each grid point received from 4 transmitter. For example, LQI values received from 4 transmitters at L_3 grid point of grid cell $B_1L_3L_4L_1$ are defined as $LQI^{B1}_{(5,0)}$, $LQI^{B2}_{(5,0)}$, $LQI^{B3}_{(5,0)}$, $LQI^{B4}_{(5,0)}$. Hence, $LQI^{k}_{(0,0)}$, $LQI^{k}_{(5,0)}$, $LQI^{k}_{(5,5)}$ and $LQI^{k}_{(0,5)}$ identify the LQI values at 4 corners of the grid cell $B_1L_3L_4L_1$ with respect to transmitters. Transmitters are defined as the additional k subscripts with LQI values at grid cell corners where k = {B_1,B_2,B_3,B_4}.

For B_1L_3 horizontal boundary of the grid cell, LQI values at virtual grid points can be interpolated in terms of LQI values at B_1 and L_3 grid points as:

$$LQI^{k}_{(i,0)} = LQI^{k}_{(0,0)} + \frac{LQI^{k}_{(5,0)} - LQI^{k}_{(0,0)}}{5} . i \qquad (4)$$

$LQI_{(i,0)}^{k}$ defines the LQI value at i^{th} virtual grid point along X axis with respect to k^{th} transmitter. Virtual LQI values along B_1L_3 can be defined for B_1transmitter by substituting $k=B_1$ and varying i between 0 and 5 in equation 4. See Table 1.

Table 1. LQI values at real and virtual grid points along B_1L_3 of $B_1L_3L_4L_1$ grid cell for B_1transmitter

Grid number (i)	LQI index	LQI values	Grid types
0	$LQI_{(0,0)}^{B1}$	$LQI_{(0,0)}^{B1}$	B_1 real grid point
1	$LQI_{(1,0)}^{B1}$	$\frac{4}{5}LQI_{(0,0)}^{B1} + \frac{1}{5}LQI_{(5,0)}^{B1}$	Virtual grid point
2	$LQI_{(2,0)}^{B1}$	$\frac{3}{5}LQI_{(0,0)}^{B1} + \frac{2}{5}LQI_{(5,0)}^{B1}$	Virtual grid point
3	$LQI_{(3,0)}^{B1}$	$\frac{2}{5}LQI_{(0,0)}^{B1} + \frac{3}{5}LQI_{(5,0)}^{B1}$	Virtual grid point
4	$LQI_{(4,0)}^{B1}$	$\frac{1}{5}LQI_{(0,0)}^{B1} + \frac{4}{5}LQI_{(5,0)}^{B1}$	Virtual grid point
5	$LQI_{(5,0)}^{B1}$	$LQI_{(5,0)}^{B1}$	L_3real grid point

By varying k values, virtual LQI values with respect to other transmitters can also be determined along B_1L_3boundary. Hence, there are 4 virtual LQI values for 4 transmitters for each i along B_1L_3.SimilarlyFor B_1L_1 vertical boundary of the same grid cell, LQI values at the virtual grid points can be interpolated in terms of LQI values at B_1 and L_1 grid points as:

$$LQI_{(0,j)}^{k} = LQI_{(0,0)}^{k} + \frac{LQI_{(0,5)}^{k} - LQI_{(0,0)}^{k}}{5}.j \qquad (5)$$

j is the number of virtual grid points along Y axis and k represents transmitters. $LQI_{(0,j)}^{k}$ defines the LQI value at j^{th} virtual grid point with respect to k^{th} transmitter. Virtual LQI values along B_1L_1 can be defined for B_1transmitter by substituting $k=B_1$ and j varies between 0 and 5 in equation 5. See Table 2.

Table 2. LQI values at real and virtual grid points along B_1L_1 of $B_1L_3L_4L_1$ grid cell for B_1transmitter

Grid number (j)	LQI index	LQI values	Grid types
0	$LQI_{(0,0)}^{B1}$	$LQI_{(0,0)}^{B1}$	B_1 real grid point
1	$LQI_{(0,1)}^{B1}$	$\frac{4}{5}LQI_{(0,0)}^{B1} + \frac{1}{5}LQI_{(0,5)}^{B1}$	Virtual grid point
2	$LQI_{(0,2)}^{B1}$	$\frac{3}{5}LQI_{(0,0)}^{B1} + \frac{2}{5}LQI_{(0,5)}^{B1}$	Virtual grid point
3	$LQI_{(0,3)}^{B1}$	$\frac{2}{5}LQI_{(0,0)}^{B1} + \frac{3}{5}LQI_{(0,5)}^{B1}$	Virtual grid point
4	$LQI_{(0,4)}^{B1}$	$\frac{1}{5}LQI_{(0,0)}^{B1} + \frac{4}{5}LQI_{(0,5)}^{B1}$	Virtual grid point
5	$LQI_{(0,5)}^{B1}$	$LQI_{(0,5)}^{B1}$	L_3 real grid point

k values can be varied again and virtual LQI values with respect to other transmitters can be determined similarly along B_1L_1 boundary. A schematical representation of linear LQI distributions along B_1L_3 and B_1L_1 boundaries of the grid cell $B_1L_3L_4L_1$ for transmitter B_1, is given in Figure 4.

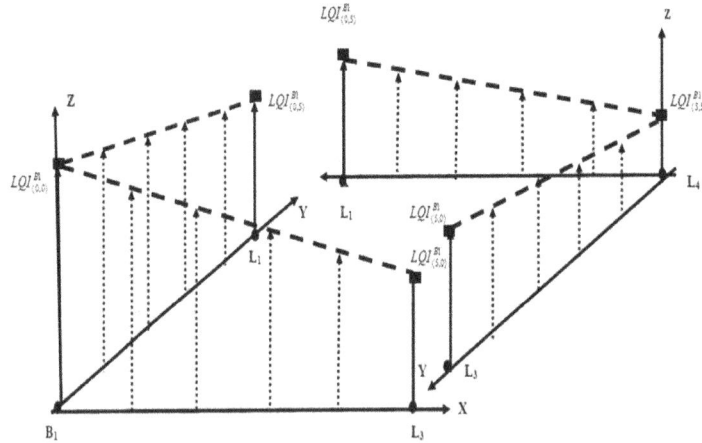

Figure 4. Linear LQI distributions are displayed along the grid cell boundaries B_1L_3 and B_1L_1 for B_1 transmitter. Each dotted arrow represents a virtual LQI value between grid points

Virtual LQI values at boundaries L_1L_4 and L_3L_4 of the grid cell $B_1L_3L_4L_1$ can also be determined for B_1 transmitter similar to B_1L_3 and B_1L_1. Once virtual LQI values are determined for all the grid cell boundaries, virtual LQI values at virtual grid points inside the grid cell can be calculated with respect to these boundary values. For j=1 and i varies between 0 and 5, virtual LQI values can be determined for kth transmitter by using the following LQIboundary values.

$$LQI_{(0,0)}^k + \frac{1}{5}(LQI_{(0,5)}^k - LQI_{(0,0)}^k) \text{ and } LQI_{(5,0)}^k + \frac{1}{5}(LQI_{(5,5)}^k - LQI_{(5,0)}^k)$$

Similarly, for i=1 and j varies between 0 and 5, virtual LQI values can also be determined by using the following LQI boundary values.

$$LQI_{(0,0)}^k + \frac{1}{5}(LQI_{(5,0)}^k - LQI_{(0,0)}^k) \text{ and } LQI_{(0,5)}^k + \frac{1}{5}(LQI_{(5,5)}^k - LQI_{(0,5)}^k)$$

These boundary LQI values can be displayed with respect to grid cell in Figure5.

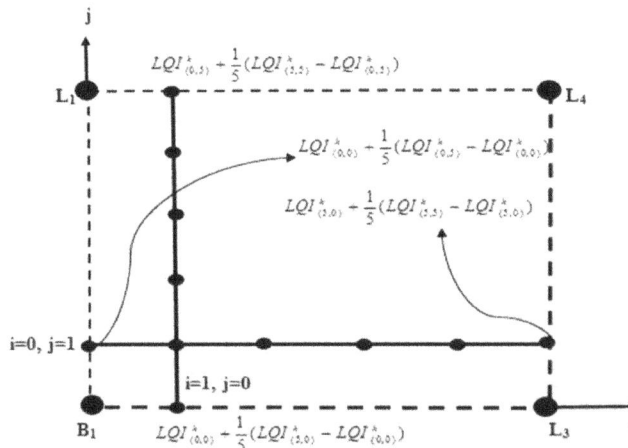

Figure 5. LQI boundary values for j=1, i = 0 to 5 and i=1, j= 0 to 5 for k^{th} transmitter are displayed

When j=1 and i varies between 0 and 5 along the horizontal line parallel to B_1L_3 , virtual LQI values can be shown in Table 3 for B_1 transmitter . Similarly, when i=1 and j varies between 0 and 5 along the vertical line parallel to B_1L_1 ,virtual LQI values can also be shown in Table 4 for B_1 transmitter.

These Virtual LQI calculations will be repeated for each horizontal and vertical line along i and j directions in the grid cell. As a result, there will be 2 virtual LQI values generated for each virtual grid point (i, j) . Average of two virtual LQI values is taken as the final virtual LQI value at that inner virtual grid point (i, j) for k transmitter.

Table 3. LQI values for j=1, i=0 to 5 for B_1 transmitter in $B_1L_3L_4L_1$ grid cell

Grid No (i , j)	LQI values(horizontal to B_1L_3)
0 , 1	$\frac{4}{5} LQI\,^{B1}_{(0,0)} + \frac{1}{5} LQI\,^{B1}_{(0,5)}$
1 , 1	$\frac{16}{25} LQI\,^{B1}_{(0,0)} + \frac{4}{25} LQI\,^{B1}_{(0,5)} + \frac{4}{25} LQI\,^{B1}_{(5,0)} + \frac{1}{25} LQI\,^{B1}_{(5,5)}$
2 , 1	$\frac{12}{25} LQI\,^{B1}_{(0,0)} + \frac{3}{25} LQI\,^{B1}_{(0,5)} + \frac{8}{25} LQI\,^{B1}_{(5,0)} + \frac{2}{25} LQI\,^{B1}_{(5,5)}$
3 , 1	$\frac{8}{25} LQI\,^{B1}_{(0,0)} + \frac{2}{25} LQI\,^{B1}_{(0,5)} + \frac{12}{25} LQI\,^{B1}_{(5,0)} + \frac{3}{25} LQI\,^{B1}_{(5,5)}$
4 , 1	$\frac{4}{25} LQI\,^{B1}_{(0,0)} + \frac{1}{25} LQI\,^{B1}_{(0,5)} + \frac{16}{25} LQI\,^{B1}_{(5,0)} + \frac{4}{25} LQI\,^{B1}_{(5,5)}$
5 , 1	$\frac{4}{5} LQI\,^{B1}_{(5,0)} + \frac{1}{5} LQI\,^{B1}_{(5,5)}$

Table 4. LQI values for i=1, j=0 to 5 for B_1 transmitter in $B_1L_3L_4L_1$ grid cell

Grid No(i , j)	LQI values(horizontal to B_1L_4)
1 , 0	$\frac{4}{5} LQI\,^{B1}_{(0,0)} + \frac{1}{5} LQI\,^{B1}_{(5,0)}$
1 , 1	$\frac{16}{25} LQI\,^{B1}_{(0,0)} + \frac{4}{25} LQI\,^{B1}_{(5,0)} + \frac{4}{25} LQI\,^{B1}_{(0,5)} + \frac{1}{25} LQI\,^{B1}_{(5,5)}$
1 , 2	$\frac{12}{25} LQI\,^{B1}_{(0,0)} + \frac{3}{25} LQI\,^{B1}_{(5,0)} + \frac{8}{25} LQI\,^{B1}_{(0,5)} + \frac{2}{25} LQI\,^{B1}_{(5,5)}$
1 , 3	$\frac{8}{25} LQI\,^{B1}_{(0,0)} + \frac{2}{25} LQI\,^{B1}_{(5,0)} + \frac{12}{25} LQI\,^{B1}_{(0,5)} + \frac{3}{25} LQI\,^{B1}_{(5,5)}$
1 , 4	$\frac{4}{25} LQI\,^{B1}_{(0,0)} + \frac{1}{25} LQI\,^{B1}_{(0,5)} + \frac{16}{25} LQI\,^{B1}_{(5,0)} + \frac{4}{25} LQI\,^{B1}_{(5,5)}$
1 , 5	$\frac{4}{5} LQI\,^{B1}_{(0,5)} + \frac{1}{5} LQI\,^{B1}_{(5,5)}$

3.2.Hybrid Exponential Interpolation Function

Transmitted LQI values decrease with respect to distance between the transmitter and a wireless sensor receiver. A best fit curve on the experimental LQI distribution canbe shown as exponential function in the form of ae^{-bx} . This is presented in implementation section.

Due to the exponential decreasing propertiesof LQIvalues between transmitters and receivers across the grid space, LQI values received at grid points are also assumed to be exponentially decreasing along the directions of transmissions at thegrid cell boundaries. These assumptions are presented schematically in Figure 6and Figure 7for B_1transmitter along the boundaries ofB_1,L_3,L_4,L_1 grid cell.

Coordinates of the virtual grid points are calculated with respect to transmitter coordinates. There are k number of LQI values arrivingfrom ktransmitters recordedat each physical grid pointto generate virtualexponential LQI distributionsbetween 2adjacent grid points along the cell boundaries.

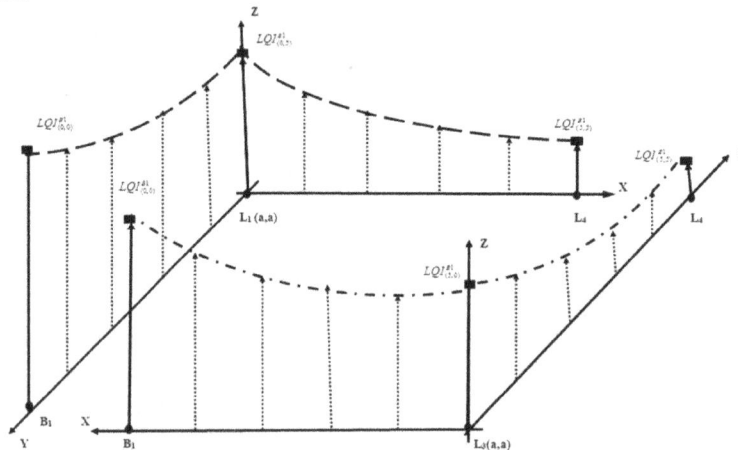

Figure 6. Schematical view of Exponential LQI interpolation functions starting from grid points a) B_1 and b) L_4 along the cell boundaries for LQI transmissions of B_1 transmitter

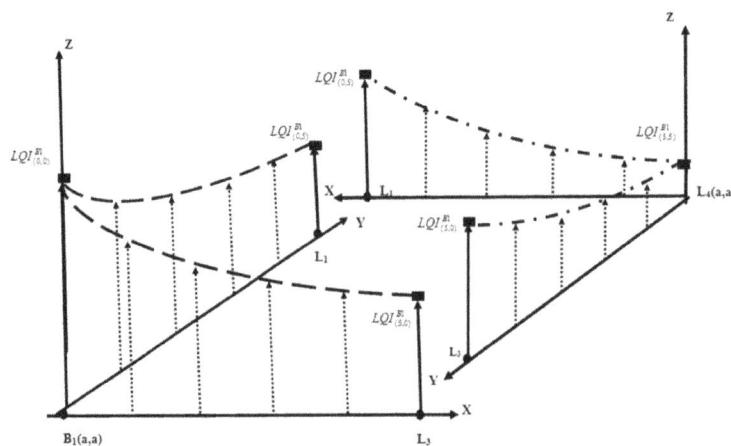

Figure 7. Schematical view of Exponential LQI interpolation functions starting from grid points a) L_1 and b)L_3 along the cell boundaries for LQI transmissions of B_1 transmitter

Grid cell $B_1L_3L_4L_1$ is considered for the realization of virtual LQI values at virtual grid points as an example. LQI value received from B_1 transmitter at grid point B_1 is identified as $LQI_{(0,0)}^{B1}$ and it decreases exponentially towards L_3 and L_1 adjacent grid points. Similarly LQI value received from B_2 transmitter at grid point L_3 is identified as $LQI_{(5,0)}^{B2}$ and this value decreases exponentially towards B_1 and L_4 grid points.

The decrease of LQI values isidentified with exponential taper function between two LQI values coming from the same transmitter at 2 adjacent grid points. The taper function for LQI values,transmitted from B_1 transmitter,between B_1 and L_3 is shown in Figure 8 and expressed as;

$$(LQI_{(a,a)}^{B1} - LQI_{(a+5,a)}^{B1})e^{-(x-a)} + LQI_{(a+5)}^{B1}$$

On the other hand, the taper function for LQI values between B_1 and L_3,transmittedfrom B_2transmitter,is shownin Figure 9 andexpressed as;

$$(LQI_{(a+5,0)}^{B2} - LQI_{(a,a)}^{B2})e^{-(x-a-5)} + LQI_{(a,a)}^{B2}$$

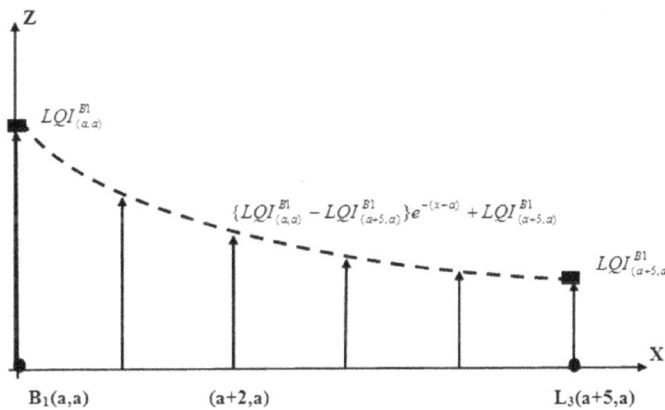

Figure 8. Graphical view of exponential taper function for grid cell boundary B_1L_3 and 4 virtual grid points with B_1 transmission

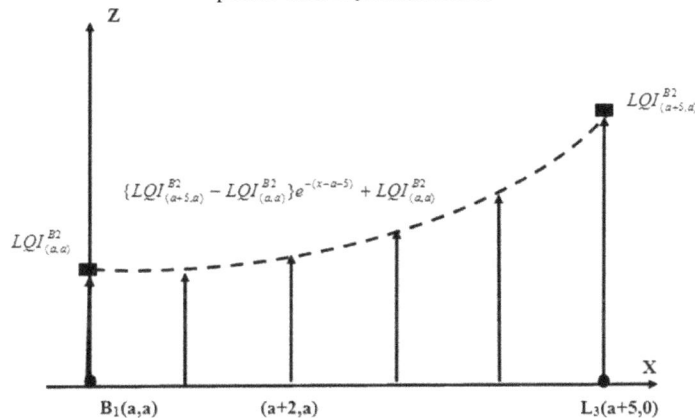

Figure 9. Graphical view of exponential taper function for grid cell boundary B_1L_3 and 4 virtual grid points with B_2 transmission

Virtual LQI values along B_1L_3, B_1L_1, L_1L_4 and L_3L_4 boundaries can be calculated by using exponential taper functions as shown in Figure 6 and7. These virtual LQI values are tabulated in Table 5.Each grid corner B_1,L_1,L_3 and $L_4$is considered as theRF transmission coordinatecenter (a,a) during calculations.Virtual grid points are located at incremental steps of 1along the grid boundaries with a generalcoordinate system of origin (a,a) corresponding to origin (0,0).

Table 5. LQI values at real and virtual gridpoints around the grid cell boundaries for B_1transmissions

$$\text{where } A = \{LQI^{B1}_{(a,a)} - LQI^{B1}_{(a+5,a)}\}, B = LQI^{B1}_{(a+5,a)}, C = \{LQI^{B1}_{(a,a)} - LQI^{B1}_{(a,a+5)}\},$$

$$D = LQI^{B1}_{(a,a+5)}, E = \{LQI^{B1}_{(a,a+5)} - LQI^{B1}_{(a+5,a+5)}\}, F = LQI^{B1}_{(a+5,a+5)}, R = LQI^{B1}_{(a+5,a+5)},$$

$$P = \{LQI^{B1}_{(a+5,a)} - LQI^{B1}_{(a+5,a+5)}\},$$

Grid locations along X	LQI values B_1L_3 Boundary	LQI values L_1L_4 Boundary	Grid locations along Y	LQI values B_1L_1 Boundary	LQI values L_3L_4 Boundary
a	$LQI^{B1}_{(a,a)}$	$LQI^{B1}_{(a,a+5)}$	a	$LQI^{B1}_{(a,a)}$	$LQI^{B1}_{(a+5,a)}$
a+1	$Ae^{-1} + B$	$Ee^{-1} + F$	a+1	$Ce^{-1} + D$	$Pe^{-1} + R$
a+2	$Ae^{-2} + B$	$Ee^{-2} + F$	a+2	$Ce^{-2} + D$	$Pe^{-2} + R$
a+3	$Ae^{-3} + B$	$Ee^{-3} + F$	a+3	$Ce^{-3} + D$	$Pe^{-3} + R$
a+4	$Ae^{-4} + B$	$Ee^{-4} + F$	a+4	$Ce^{-4} + D$	$Pe^{-4} + R$
a+5	$Ae^{-5} + B$	$Ee^{-5} + F$	a+5	$Ce^{-5} + D$	$Pe^{-5} + R$

Once the virtual LQI values are determined for one transmitter around the grid cell boundaries, other virtual LQI values can also be determined for other transmitters around the same boundaries. Boundary LQI values of a grid cell are utilized to calculate the internal virtual LQI values of the grid cell.

Virtual LQI values on two cell boundaries facing opposite to each other are considered and the virtual LQI values are calculated betweenthem by using linear interpolation technique. Final virtual LQI value is derived by averaging the two resultant LQI values obtained horizantal and vertical directions.

4. IMPLEMENTATION

JENNIC JN5139 wireless sensor nodes are deployed in the study. Zigbee Home Sensor program is used to program JN5139 active devices to work as both transmitter and receiver WSNs respectively [14]. JN5139 receiver,[15],on the object is interfaced to a computer via a wireless link for data transmission.ZigBee protocol which is based on IEEE 802.15.4 standard in 2.4 GHz frequency band is used during the communication and data transmission between the transmitter and receiver nodes.

A rectangular area of 20m x 12m in a sports hall is selected and unknown object locations are limited in this rectangular grid area.The area was not free of obstacles. There were sportsequipments lying around and people were doing sports and moving around during the measurements. Wireless Sensor transmitters are placed at the corners of the rectangular area.Recordings of LQI values coming from transmitters are collected by a wireless sensor receiversequentially placed at each grid point. Power consumption by the sensors during the construction of fingerprint map and computations is negligible. The wireless sensors are active devices and their onboard battery life is around 1 month. Total LQImeasurements, datacollection and recording in fingerprint database takes only 1-2 hours. Secondly, there is no onboard processor and as a result, there are no onboard computations with these sensors. Construction of the database and the localization computations take place in server computer. Hence the only energy used by the sensors is to transmit the LQI values.

For a grid area of 20mx12m, 24 grid points are arranged with a grid space of 4 meters. There are 96 LQI entries recorded in the fingerprint database with 4 LQI readings at each grid point from 4 transmitters. Each entry in the database includes a mapping of the grid coordinate (x,y) and 4 LQI values at that point.

Wireless receiver on the object receives 4 LQI values from 4 transmitters from its 4 channels and transmits them to server computer via a wireless link. There is no onboard memory at the receiver and these values are stored sequentially in an access database in a servercomputer.Server computer has sufficient memory space to manipulate these LQI values for position calculations.Received signal strengths can vary depending on the environmental effects. These variations are reduced by averaging 100recorded LQI values at each measurement pointfor each transmitter. Averaged LQI values and the position coordinates are employed to generate the fingerprint map in the server.LQI recordings of the object receiver at unknown locations are also carried outsimilarly to generate object fingerprint vectors.

4.1.Linear Interpolation

Each Grid cell is further divided into 5x5 virtual grid cells as an example. Therefore there are total 3x5x25=375 grid cellsand 26x16 =416 grid points and 416 LQI values from each transmitters across the testing area.

Initially, k-NN and weighted k-NN algorithms are utilized to determine the unknown target locations by using basic fingerprint database with 24 grid points. Results are presented in Table 6. The same localization algorithms are deployed with virtual fingerprint database generated with Linear interpolation function. Unknown object coordinates are determined and the results are presented in Table 6.

Table 6.Estimated object position coordinates using basic fingerprint database and virtual fingerprint databases with linear and hybrid exponential interpolation functions

| Unknown object | Estimated object position coordinates   using k-NN and weighted k-NN | | | | | | | | | | | | | | |
| | 1-NN X,Y | | | 2-NN X,Y | | | 3-NN X,Y | | | 4-NN X,Y | | | Weighted 4-NN X,Y | | |
X Y	Basic finger print	Linear taper	Expo. taper	Basic finger print	Linear taper	Expo. taper	Basic finger print	Linear taper	Expo. taper	Basic finger print	Linear taper	Expo. taper	Basic finger print	Linear taper	Expo. taper
2 2	1 3	0 3.8	0.9 1.3	0 3	0.3 3.5	1.2 1.4	1 2	0.3 3.7	1.0 0.9	0 4	0.4 4.1	1.3 0.9	0.9 2.9	0.5 3.6	0.9 1.3
2 3	1 4	0 5.5	1.2 1.3	1 5	0.2 4.5	0.8 1.4	0 3	0 4.7	1.3 1.2	0 4	0.1 5.2	1.2 1.5	0.5 4.5	0.2 4.5	1.4 1.6
3 5	2 4	1.1 8.0	2.2 3.5	2 3	1.3 7.6	1.8 3.7	2 6	1.2 7.1	1.7 4.1	2 3	0.9 9.0	2.6 2.7	2.1 3.3	1.4 8.6	2.1 2.8
4 4	0 12	5.8 3.2	4.5 2.4	4 8	5.2 2.4	5.3 2.4	7 9	5.4 2.4	6.4 3.1	6 7	5.3 1.6	3.2 2.2	5.8 7.31	5.8 2.0	5.3 3.2
0 8	4 12	2.4 10	1.7 8.6	4 6	2.4 9.8	13 9.6	5 5	2.4 9.2	1.6 9.3	4 7	1.4 9.8	1.4 10.0	4.14 7.5	2.5 9	1.7 9.1
8 8	8 12	8 6.4	7.1 6.7	8 12	8 6.6	6.5 6.6	8 9	7.2 6.4	6.6 6.8	9 9	7.2 7.1	6.8 6.7	9.8 9.2	7.2 7.4	7.3 6.2
12 8	12 12	11 6.6	11 7.1	12 10	10.5 6.8	10.7 6.9	11 11	10.4 7.1	10.4 6.7	10 8	10.5 6.7	11.2 6.9	10.3 8.8	10.6 6.8	10.4 6.7
4 12	8 8	5.6	4.8 10.8	6 12	4.6 13.2	5.5 10.7	5 11	4.8 13.2	5.4 13.4	6 11	5.6 13.2	5.6 11.6	6.36 10.2	5.6 13.6	4.6 13.7
8 12	8 8	6.4 14	7.1 10.6	6 8	7.4 13	6.8 10.7	5 11	6.5 13.6	6.5 10.8	5 11	6.2 14	6.2 10.8	7.01 8.8	6.6 13.8	6.4 13.8
4 16	8 12	5.4 14.4	5.0 14.7	8 10	5.6 17.6	4.7 14.7	8 12	5.7 18	5.5 17.3	7 13	4.4 16.6	5.1 14.6	7.43 12.3	5.3 17	5.7 14.0

4.2.Hybrid Exponential Interpolation

RF signal amplitudes decrease with the distance as they reach to receivers from transmitters. Generally, this decrease is in exponential form. Initially,wireless sensor receivers are placed in front of wireless sensortransmitters. The distance between them is increased in steps of 1 metre and RF signal amplitude recordingsby the receiver in the form of LQI values are plotted against distance.A best fit curve reveals an exponential distribution function of ae^{-bx} as seen in Figure 10.

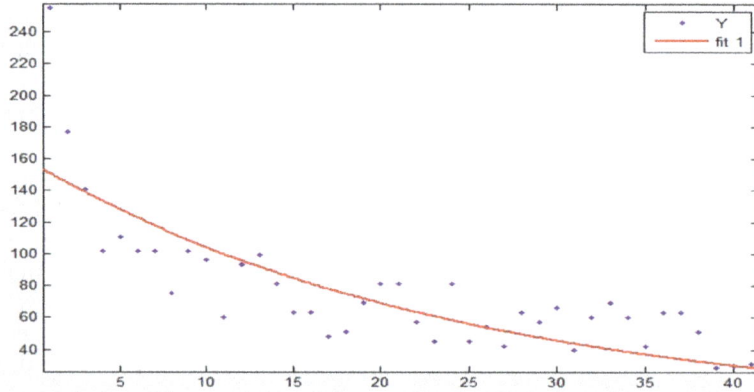

Figure 10. Plot of LQI values versus distance between a receiver and a transmitter with a best fit curve

In order to reflect this characteristic in the virtual world, Virtual Fingerprint map is generated by using exponential taper function between every 2 adjacent grid points. Initially, LQI values are recorded at each grid point. Exponential Taper functions are applied in **x**and **y** directions of the grid cell asseen in Figure 6 and 7.LQI amplitude at every virtual grid point is calculated with the assistance of exponential taper functions.k-NN andweighted k-NN localization algorithms are employed to calculate the unknown object coordinates with the virtual database generated usingExponential taper function. The results are presented in Table6.The location estimation error, e, is defined by the linear distance between the unknown object coordinates (x_t, y_t) and their estimated coordinates (x_e, y_e). It is given by:

$$e = \sqrt{(x_e - x_t)^2 + (y_e - y_t)^2}$$

 Error calculation results between the actual and the estimated object coordinates by using basic fingerprint database and the virtual fingerprint databases are tabulated in Table 7. It can be concluded that the best localization results are achieved with exponential taper function.

5. CONCLUSIONS

Virtual localization approach is a novel and time effective indoor localization technique. The proposed positioning system uses a number of transmitters and a receiver as in basic fingerprint systems. The originality lies in the introduction of virtual grid points with specific LQI taper functions among the physical grid points.

Previously, fingerprint mapping techniques are utilized to obtain localization accuracies of around 1 grid space.To increase the accuracies, number of fingerprint points are increased across the sensing area and the localization accuracies of slightly less that a grid space are achieved. This improved the accuracies but also increased the effort to build a fingerprint map. The key idea of the proposed approach is to obtain a more accurate object localization by keeping the same fingerprint map but increase the number of grid points. Each grid point is an LQI measurement point and the recordings of LQI values at these points are utilized to estimate the object location. A solution to increase the positioning accuracy is to put more grid points which will be more labour intensive and time consuming. The idea behind the proposed approach is to simulate a larger number of grid pointsby introducing virtual grid points and keeping the samenumber of real grid points.

Table 7. Overall error calculations for 3 fingerprint database systems.

Unknown positions	Error calculations between target and average estimated positions using basic and virtual fingerprint databases (linear and hybrid exponential interpolation functions)					
X Y	Basic fingerprint ave	Error (m)	Virtual Fingerprint Ave (linear taper)	Error (m)	Virtual Fingerprint Ave (exp. taper)	Error (m)
2 2	0.5 3	1.8	0.3 3.8	2.5	1.0 1.2	1.3
2 3	0.5 4.1	1.9	0.1 4.9	2.7	1.2 1.4	1.8
3 5	1.5 2.8	2.7	1.2 8.0	3.5	2.0 3.3	2.0
4 4	4.5 8.6	4.6	5.5 2.3	2.3	4.9 2.6	1.6
0 8	4.2 7.5	4.2	2.2 12.1	4.7	1.5 9.3	1.5
8 8	8.5 10.2	2.2	7.5 6.8	1.3	6.8 6.6	1.8
12 8	11.1 9.9	2.0	10.6 6.8	1.8	10.7 6.8	1.8
4 12	6.3 10.4	2.8	5.2 13.1	1.6	5.3 12.2	1.3
8 12	6.2 9.4	3.1	6.6 13.6	2.1	6.5 11.3	1.7
4 16	7.6 11.8	5.5	5.2 16.7	1.4	5.6 15.0	1.9
Total Avg. error(m)		3.1		2.4		1.7

The proposed system has the following advantages. Firstly, the hardware cost is the same as fingerprint localization systems. Secondly, the number of measurement points corresponding to grid points in the test area is unchanged and only extra virtual grid points are introduced between these grid points. Hence less time and effort is spent during off line phase. Both real and virtual grid points are used together to generate a new fingerprint database for location determination.

The shortcomings of the virtual grids are their numbers across the sensing area. In theory, higher the density of grid points, greater the localization accuaracy. Although there is no extra cost of having more virtual grid points, maximum number of virtual grid points is limited with respect to localization accuracies. There is a trade off between the localization accuracies and the number of real and virtual grid points. Optimum localization accuriess are obtained with n=5. Hence, 4 virtual grid points are utilized between two adjacent real grid points by dividing the distance into n=5 equal sections between them. Any other number of virtual grid points betweentwo adjacent grid points reduced the positioning accuracies with both taper functions. In the study, total number of grid points is 416 across the test area with 24 of them are real and 392 of them are virtual. If the number of virtual grid points are increased more than 392 with n>5, there is a deterioration observed in localization accuracies. Any virtual grid number less than 392 with n<5, alsocauses a degration in localization accuracies. Consequently in our approach an optimum of 15 grid cells each with 25 virtual grid cells are employed across the sensing area by using n=5.

Once the total grid space is determined, LQI values at physical grid points are interpolated among the virtual grids corresponding to predefined taper functions. In fingerprint localization systems, fingerprint database is compared with unknown fingerprint signatures of the objects.
Basic fingerprint approach has an average localization error of 3.1m while Linear interpolation technique has an average localization error of 2.4m where LQI values are linearly distributed between the virtual grid points. Hybrid exponential interpolationtechnique, on the other hand, has an average error of 1.7m. LQI distribution between the grid points simulates the propagation characteristicsof LQI values against distance.Exponential approach gives the minimum distance error among 3 techniques.

In this study, position detection is implemented in a confined area of a sports hall. But the same technique can also be generalized in any indoor area. The main idea was to see the applicability of virtual fingerprint technique in any test area. In second stage, other indoor areas will also be tested with this newtechnique. Environmental conditions affect the LQI reception by the receivers. If there are more obstacles in the sensing area,localization accuracies are decreased accordingly, To reduce these affects, signal averaging and outlier techniques are employed on the recorded LQIvalues. Based on the observations of the complex relationship between LQI values and the distances between transmitters and receivers, Non linear interpolation algorithms can improve localization accuraciesin large indoor areas. Hybrid exponentialinterpolation algorithm is one of these algorithms whichcan compensate the nonlinear behaviour of RF signals and in return generates better localization accuracies.

REFERENCES

[1] Lionel M. NI and Yunhao , Liu Yiu Cho Lau and Abhishek   Patil ; LANDMARC: Indoor Location Sensing Using Active RFID Wireless Networks 10, 701–710, 2004

[2] R.Want , A.Hopper,V.Falcao and J.Gibbons; The activeBadgelocationsystem,ACM Transactions on Information systems Vol. 40, No. 1, pp. 91-102, January 1992

[3] R.Want, B.Schilit, N. Adams, R. Gold, D. Gold- berg, K.Petersen ,J.Ellis, M Weiser; The Parctab Ubiquitous Computing Experiment", Book Chapter: "Mobile Computing", Kluwer Publishing, Edited by Tomasz Imielinski, Chapter 2, pp 45-101, ISBN 0-7923-9697-9, February 1997.

[4] A.Ward, A.Jones, A.Hopper: A neq location technique for the active office, In IEEE personal Communication Magazine, Volume 4 no 5,pages 42-47 ,October 1997

[5] A.Harter ,A.Hopper, P.Steggles, A.Ward, P. Webster; The anatomy of a context aware application, In proceedings of the 5th annual ACM/IEEE ınternational conference on Mobile Computing and Networking,pages 59-68,August 1999

[6] P. Bahl, V.N. Padmanabhan; RADAR: An in-building RF-based user location and tracking system, in: Proceedings of IEEE INFOCOM 2000, Tel-Aviv, Israel (March 2000),

[7] J. Hightower, R. Want and G. Borriello; SpotON: An indoor 3D location sensing technology based on RFsignalstrength, UW CSE00-02-02, February 2000,

[8] Konrad Lorincz, Matt Welsh ; MoteTrack: A Robust, Decentralized Approach to RF-Based Location Tracking, Proceedings of the International Workshop on Location and Con13-RadioFrequencyIdentification (RFID) home page,

[9] Junhuai Li,Rui Qi,Yile Wang,Feng Wang; An RFID location Model based On Virtual reference tag spacemJournal of Computational Information systems 7:6 ,pp 2014-2111,2011

[10] Yiyang Zhao,Yunhao Liu,Lionel Ni ; Vire, Active RFID based Localization Using Virtual Reference Elimination,Int conference on paralel processing ,pp 56-63, 2007

[11] Kamol Kaemarungsi,prashant krishnamurthy " modelling of indoor positioning şystems based on location fingerprinting" ,IEEE info com 2004, 0-7803-8356-7

[12] Guang-yao Jin, Xiao-yi Lu,Myong Soon Park," An indoor localization mechanism using active RFID tag",proceedings of IEEE conference on sensor networks SUTC ,2006, 0-7695-2553-9

[13] Jan Blumenthal, Ralf grossmann, Frank Gola- towski "weighted centroid localization in zigbee-based sensor networks", CELISCA center for life science automation,2007

[14] http://www.jennic.com/jennic_support/application_notes/jn-an-1052_home_sensor_demons tration_using_zigbee

[15] http://www.jennic.com/support/solutions/00004

11

Optimization of Performance Metrics of LAR in Ad-Hoc network

Neelesh Gupta[1] and Roopam Gupta[2]

[1] Research Scholar, UIT, RGPV, Bhopal (M.P.)-India
neelesh.9826@gmail.com

[2] Dept. of IT, UIT, RGPV, Bhopal (M.P.)-India
roopamgupta@rgtu.net

ABSTRACT

Routing in Mobile Ad-Hoc Network (MANET) is a crucial task due to highly dynamic network environment. In latest years, several routing protocols have been implemented. In recent developments, position-based routing protocols exhibit better scalability performance and robustness against frequent topological changes such as, Location-Aided Routing (LAR) protocol. In this paper, developments of performance-metrics using LAR have been reported. Using these developments, LAR protocol has been found to be better than other protocols like DSR in MANETs of different scenario. This is because of already available location information of the nodes in network while it is based upon source routing as a DSR. Via LAR, performance metrics Packet Delivery Fraction (PDF), routing overhead, End to End (E2E) Delay, Non-Routing Load (NRL) and Number of lost (dropped) packets during route discovery can be optimized in dynamic Ad-Hoc networks.

KEYWORDS

MANETs, LAR, DSR, Performance Metrics.

1. INTRODUCTION

Mobile Ad-Hoc Network is an infrastructure less, self-organizing, self-configuring, self-maintaining network designed by a set of wireless mobile nodes, where all the mobile hosts take part in the process of forwarding packets. These are highly applicable in Military Networks, Personal Area Networks, Home Networks, Wireless Sensor Networks, and Inter-Vehicle Communication. Each node in the network also acts as a router, forwarding data packets for other nodes. A central challenge in the design of Ad-Hoc networks is the development of dynamic routing protocols. A routing protocol is needed whenever a packet needs to be transmitted to a destination via number of nodes and numerous routing protocols have been proposed for Ad-Hoc networks. Several routing protocols have been planned to achieve a particular level of routing operation for MANET. The routing protocols are divided into several categories. The most popular classification is between Topological and Position-based routing Protocols. Under Topological-based proactive and reactive protocols come. In proactive a source node wants to transmit the data from S to D, it searches the routing table to find a destination node match. Destination Sequenced Distance Vector (DSDV), Wireless Routing Protocol (WRP), Cluster Switch Gateway Routing (CGSR), Source Tree Adaptive Routing Protocol (STAR) is the examples of proactive protocols. The main drawback of these protocols is that the maintenance of unused paths may occupy a significant part of the available bandwidth if the topology of the network changes frequently. In reactive, the routes are discovered only when the source needs to transmit the data. Dynamic Source Routing Protocol (DSR), Ad-Hoc On-Demand Distance-Vector Routing Protocol (AODV), Temporally-Ordered

Routing Algorithm (TORA) is the reactive routing protocols [7] [9]. These reduce the burden on the network when only a small subset of all available routes is in use at any time. Further position-based routing protocols eliminate some of the inherent limitations of topology-based routing by using additional location information. The routing decision at each node is then based on the destination's position contained in the packet and the position of the forwarding node's neighbours. Position-based routing does not require the establishment or maintenance of routes [9]. There are three main packet forwarding schemes in position-based routing: Greedy-based, Restricted-directional flooding and Hierarchical approaches. For the first two, a MH forwards a given packet to one (greedy-forwarding) or more (restricted directional flooding) one-hop neighbours. The third forwarding strategy forms a hierarchy in order to scale to a large number of MHs. Greedy-based Routing Protocols do not establish and maintain paths from S to D. In this packet sender node includes approximate position of the recipient in packet. MFR, GPSR are examples. In Restricted Directional flooding-based Routing Protocols source floods data packets in a restricted geographical area towards the direction of destination. Instead of selecting a single node as the next hop several nodes participate in forwarding the packet in order to increase the probability of finding the shortest path. Examples are Location-Aided Routing Protocol (LAR), Distance Routing Effect Algorithm for Mobility (DREAM) routing protocols. Hierarchal-based Routing Protocols form a hierarchy in order to scale to a large number of MHs. Complexity of the routing algorithm can be reduced tremendously by establishing some form of hierarchy also. Examples are GRID & TERMINODES.

LAR is source routing protocol, as a DSR. Initially it starts flooding in all the directions by the source after expecting the destination the routing is so easy in network that will be in the direction of the destination. This is the basic principle of directional-restricted position based location-aided routing in MANET. In this paper, optimizations of Ad-Hoc network performance metrics are revealed via restricted directional flooding-based LAR scheme. In the second section, related work is presented. Overview of DSR and LAR protocols is shown in the next section. In the fourth section simulation results on DSR and LAR performance discussion and displays are shown. The last section concludes the paper.

2. RELATED WORK

Earlier several researchers had been devoted their contribution for the evaluation of various routing protocols. Das S. et al. [3] compared performance of two on-demand routing protocols (AODV & DSR) for Ad-Hoc networks. The general observation from the simulation is for the application oriented metrics such as delay and throughput. DSR outperforms than AODV in less stressful situations (smaller number of nodes and lower load/mobility). However, AODV outperforms than DSR in more stressful situations (more load and higher mobility). However, DSR consistently generates less routing load than AODV. Mauve M. et al. [9] presented an overview of MANET routing protocols that make forwarding decisions based on the geographical position of a destination packet. They also provided a qualitative comparison of position-based like a DREAM, LAR, GLS and Greedy routing approaches. They also finally concluded that LAR and DREAM scheme can be used where small number of packets to be need transmitted very reliably in the network.

E.Ahver et al. [6] compared the performance of three routing protocols for Ad-Hoc network: DSR, AODV and LAR-1. Their evaluation is based on energy consumption in MANETs. Performance analysis is examined using varying network load, mobility and network size of the network. DSR consumes the least energy for low density networks. LAR-1 for high density networks is much better than others. Therefore, LAR-1 is a good protocol for high density Ad-Hoc networks. Y.B. Ko. et al. [8] proposed an optimization to route-discovery known as LAR protocol that uses GPS for location of all mobile nodes in networks. They suggested LAR approach to utilize location information using GPS to improve performance of routing protocols for Ad-Hoc networks. LAR protocols limit the search for a new route to a smaller "request zone" of the Ad-Hoc network. This results in a significant reduction in the number of routing messages and also presented two LAR algorithms to determine the request zone optimizations suggestions to algorithms. Qabajeh L. et al. [11] explained a qualitative comparison of the existing geographic routing protocols that make forwarding decisions based on the geographical position of a packets destination. Advantages of all position-based strategy are illustrated in this research paper. Finally they identified a number of research opportunities which could lead to further improvements in position-based Ad-Hoc routing also. Camp T. et al. [2] clarified the performances of LAR and DREAM routing protocols and compared the both with DSR in MANETs. Their analysis produces that via adding location information to DSR (like LAR) network load and PDF both are improved. The project aim of David D. et al. [4] was to test routing performance of four different routing protocols (AODV, DSR, LAR 1 and ZRP) as a function of network and area size. This describes the different routing protocols, the experiment setup and finally presents the simulation results. AODV outperforms than DSR. LAR-1 is even better than AODV up to 200 nodes in terms of delivery ratio and routing overhead due to the geographical information of the node. Despite the popularity of several protocols, research efforts have not focused in evaluating their performance in large-scale wireless networks. This greatly affects the network efficiency, since it necessitates frequent exchange of routing information. Broustic I. et al. [1] presented the behaviour of the DSR, AODV, TORA and LAR protocols in large-scale MANETs. They concluded that DSR scales well in terms of packet delivery fraction but suffers an important increase of end-to-end delay, as compared to its performance achieved in small scale topologies. LAR appears to scale very well in terms of all metrics employed.

3. OUTLINE OF DSR AND LAR ROUTING PROTOCOL

DSR: The Dynamic Source Routing (DSR) is one of the examples of an on-demand (reactive) routing protocol that is based on the concept of source routing. It is designed especially for use in multi-hop Ad-Hoc networks of mobile nodes. DSR is composed of the two mechanisms of route-discovery and route maintenance, which work together to allow nodes to discover and maintain source routes to arbitrary destinations in the network. When a node in the Ad-Hoc network attempts to send a data packet to a destination for which it does not know the route, it uses a route discovery process to dynamically determine a route. Route discovery works by flooding the network with route request (RREQ) packets. Each node receiving an RREQ rebroadcasts it, unless it is the destination or it has a route to the destination in its route cache. Such a node replies to the RREQ with a route reply (RREP) packet that is routed back to the original source [5] [6].

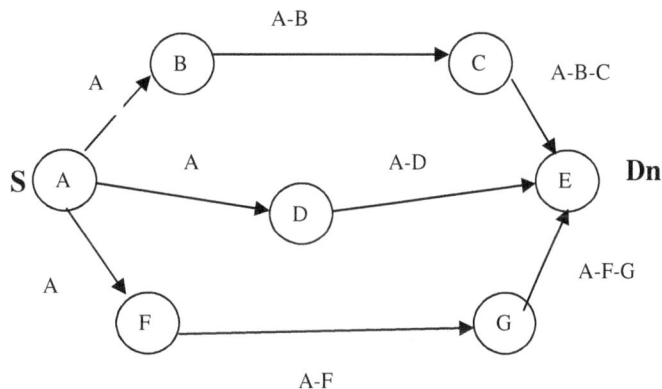

Figure.1.a RREQ Packet Route Discovery Mechanism

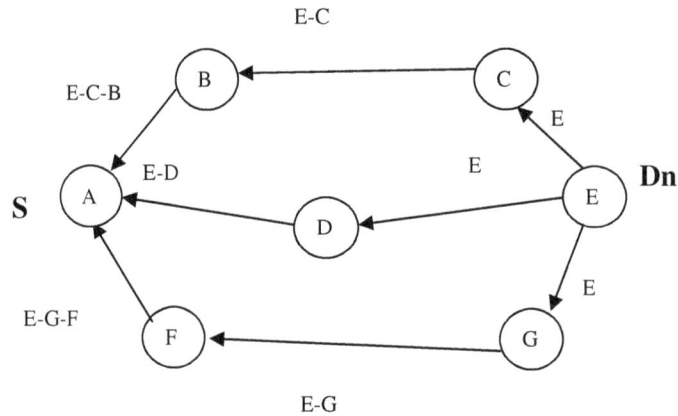

Figure.1.b RREP Packet Route Discovery Mechanism

Figure1. DSR Route Discovery Mechanism

LAR: LAR is source routing protocol, as a DSR. Initially it starts flooding in all the directions by the source after expecting the destination the routing is so easy in network that will be only in the direction of the destination. It sends the location information in all packets to decrease the routing overhead in future route discovery process in the network. It uses the location information by using GPS. Less routing overhead with LAR can be achievable by limiting the search space for the desired route to a destination into minor request regions in the network. LAR uses a request zone that is rectangular in shape. Consider a node S (Source) that needs to find a route to node D (Destination). Assume that node S knows that node D was at location (Xd, Yd) at time t_0. At time t_1, node S initiates a new route discovery for destination D. It assumes that node S also knows the average speed v with which D can move. Using this, node S defines the expected zone at time t_1 to be the circle of radius $R = v(t_1 - t_0)$ centered at location (Xd, Yd). In figure 2.a, $t_1 - t_0$ is the elapsed time between two successive route requests from the source node. When a node receives a route request, it discards the request if the node is not within the request region. For instance, in Fig.2.b, if node M receives the route request from another node, node M forwards the request to its neighbours, because M is within the rectangular request zone. However, when node N receives the route request, node N discards the request, as node N is not within the request zone [6], [7] and [9].

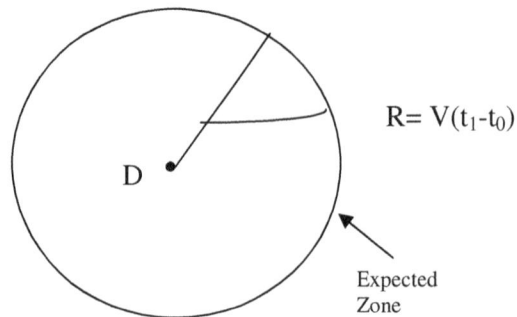

Figure2.a Expected Zone

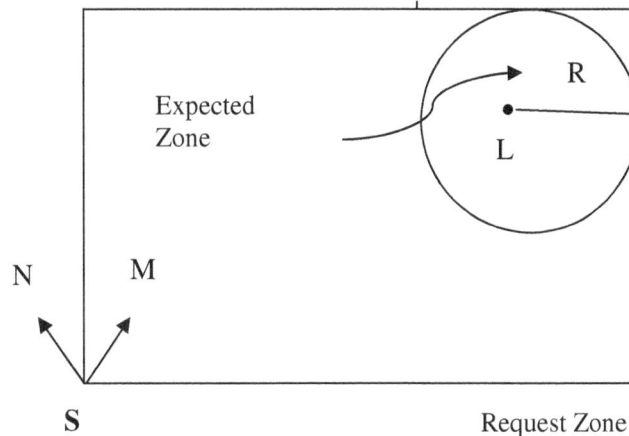

Figure 2.b LAR-1 Request Region

Figure 2. LAR-1 Scheme

DSR and LAR both use source routing in network but difference is that LAR at first start routing as a DSR but after expecting the destination packets are flooded in the direction of destination only. As a result the routing overhead is to be reduced and better performance of LAR protocol is obtained.

4. SIMULATION RESULTS AND DISCUSSION

The simulation study was conducted in the "Network Simulator" (NS2) environment and used the Ad-Hoc networking extensions provided by CMU [10]. All simulations were performed on Intel (R) core (TM) i3 CPU, 2.3 GHZ, 3072 MB of RAM running on Inspiron N5010 configuration. These include physical, data link and medium access control layer models. The Distributed Coordination Function (DCF) of IEEE 802.11 is used to model the contention of nodes for the wireless medium. The radio model uses characteristics similar to Lucent's WaveLAN direct sequence spread spectrum radio. The source-destination pairs were spread randomly over the network. Constant bit rate (CBR) traffic sources are used in simulation. The size of these packets is 512 bytes with transport agents TCP & UDP. The random waypoint mobility model has been used in a rectangular filed area with different number of mobile nodes. This model based on random waypoints and random speeds that includes pause times between changes in destination and speed. In this model, a mobile node moves from its current location to a randomly chosen new location within the simulation area using a random speed uniformly distributed between the maximum and minimum speed of the simulation. The simulations were run for 100 seconds. The number of nodes (n) in simulated Ad-Hoc network is 12, 24, 36, 48 and situated in a 800x600 square meter region having a transmission range of 550m. In this simulation, each node starts its journey from a random location to a random destination with a randomly chosen speed (uniformly distributed average speed of 20 m/sec). Once the destination is reached, another random destination is chosen after a pause time. Pause time is taken in the interval of 20 seconds during simulation. This model is often simplified by using a uniformly distributed speed.

Table 1. Simulation Environment

Network Parameter	Value
Simulator	NS-2
Simulation time	100 Seconds
Transmission range	550 m
Node movement model	Random way point
Protocols studied	DSR and LAR
Simulation area	800 x 600
Bandwidth	2 Mb/s
Traffic type	CBR
Transport Agents	TCP and UDP
Pause Time	0 - 100 s in steps of 20s
Average Node Speed	20m/s
Packet Size	512 bytes

4.1. Performance Metrics

The following performance metrics are considered for evaluation:

(a) Packet Delivery Fraction (PDF): The ratio of the data packets delivered to the destinations to packets generated by the sources. It specifies the packet loss rate which limits the maximum throughput of the network. For efficient routing protocol PDF should be more.

(b) End-to-End Delay (E2E): End-to-End Delay indicates how long time is taken for a packet to travel from the source to the destination during routing in networks. This includes all possible delays caused by buffering during route discovery latency, queuing at the interface queue, retransmission delays at the MAC, propagation and transfer times.

(c) Normalized Routing Load (NRL): The number of routing packets transmitted per data packet delivered at the destination. It is concerned with number of routing packets.

(d) Routing Overhead: The routing overhead describes how many routing packets for route discovery need to be sent in order to propagate the data packets. It is an important measure for the scalability of a protocol.

4.2. Simulation Results and Parameters

The simulation results are revealed in the following section in the form of line graphs. Graphs show comparison between the two (DSR and LAR-1) protocols by varying different numbers of nodes on the basis of the above-mentioned performance metrics as a function of pause time and number of nodes.

4.2.1 Results-

After simulating in NS-2 simulator value of PDF increases using LAR by 0.56%, reduces routing overhead or RLOAD, E2E Delay and Dropped packets by 9.25%, 17.8% and 28.6% respectively compared to DSR protocol. NRL for both is almost equal.

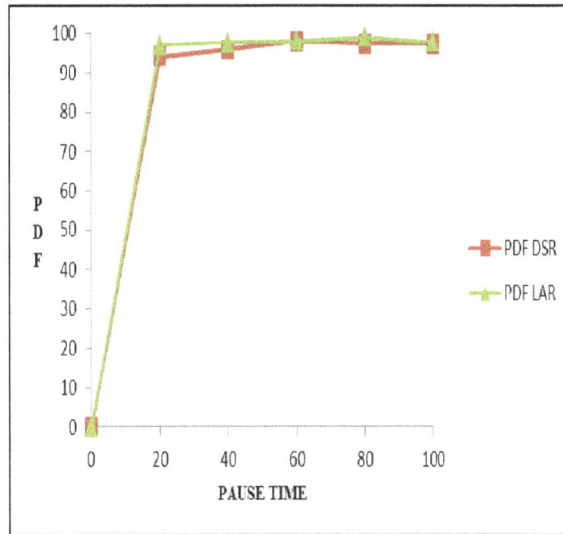

Figure.3 (a) PDF Vs Pause Time

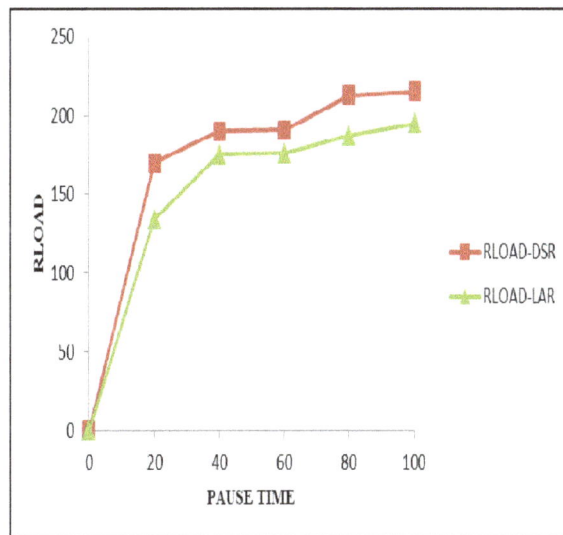

Fig.3 (b) Routing Overhead (RLOAD) Vs Pause Time

Fig.3 PDF and RLOAD variations (n=12)

4.2.2 Results-

After simulating in NS-2 value of PDF increases using LAR by 0.86%, reduces routing overhead, NRL, E2E Delay and Dropped packets by 38%, 40%, 17.5%, 30.2% respectively compared to DSR protocol.

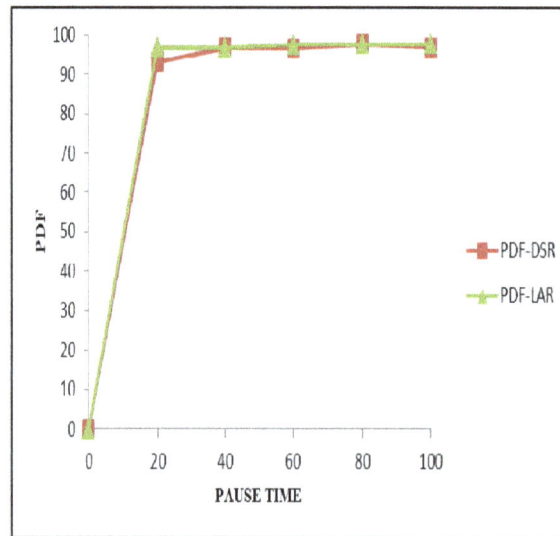

Fig.4 (a) PDF Vs Pause Time

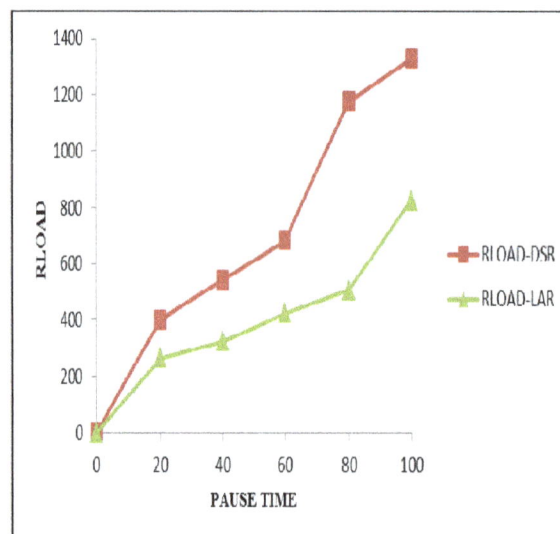

Fig.4 (b) RLOAD Vs Pause Time

Fig.4 PDF and RLOAD variations (n=24)

4.2.3Results-

After simulating in NS-2 simulator value of PDF increases using LAR by 2.93%, routing overhead, NRL, E2E Delay and Dropped packets by 22.5%, 32.5%, 47.75%, 53.28% respectively compared to DSR protocol.

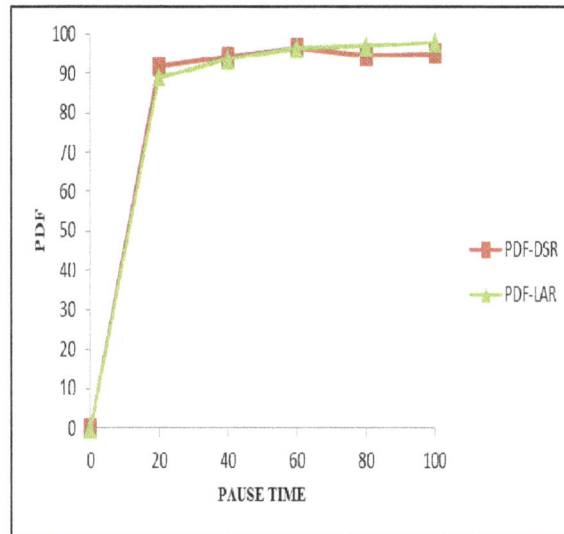

Fig.5 (a) PDF Vs Pause Time

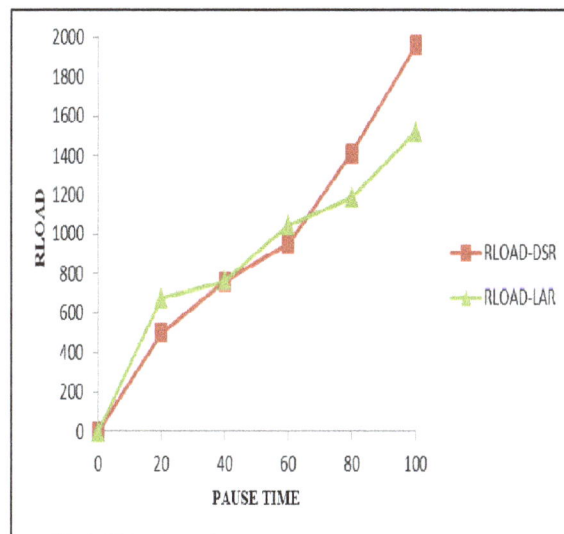

Fig.5 (b) RLOAD Vs Pause Time

Fig.5 PDF and RLOAD variations (n=36)

4.2.4 Results-

After simulating in NS-2 simulator value of PDF increases using LAR by 3.33% and reduces routing overhead, NRL, E2E Delay and Dropped packets by 3.25%, 10.9%, 26.68%, 9.1% respectively compared to DSR protocol.

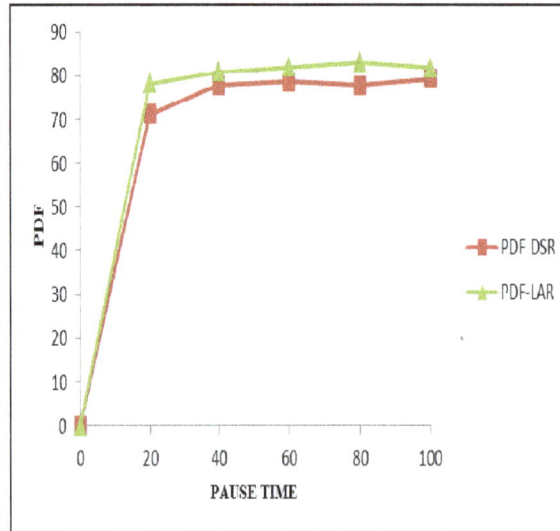

Figure.6 (a) PDF Vs Pause Time

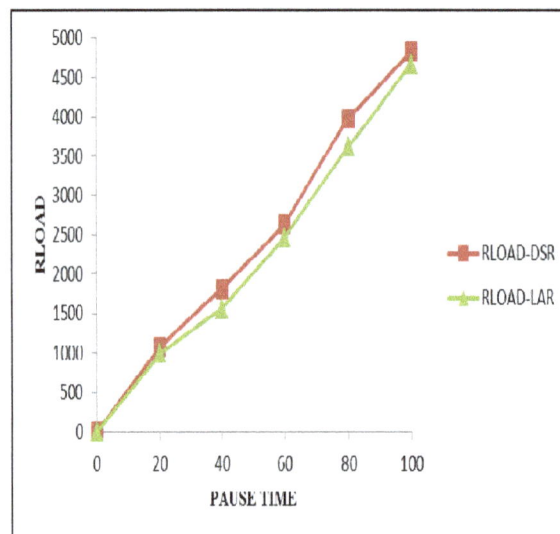

Figure.6 (b) RLOAD Vs Pause Time

Fig.6 PDF and RLOAD variations (n=48)

4.2.5 Results-

Next graph results are plotted between PDF versus Number of nodes and Routing Overhead versus Number of nodes. As the number of nodes increase into the network, PDF ratio using LAR increases in a better ratio compared to DSR protocol. Similarly routing overhead value using LAR reduces more compared to DSR protocol as the number of nodes increases into the network. Because of reduced routing overhead NRL and E2E is to be reduced. Along with optimization of these performance metrics number of lost packet is also to be reduced.

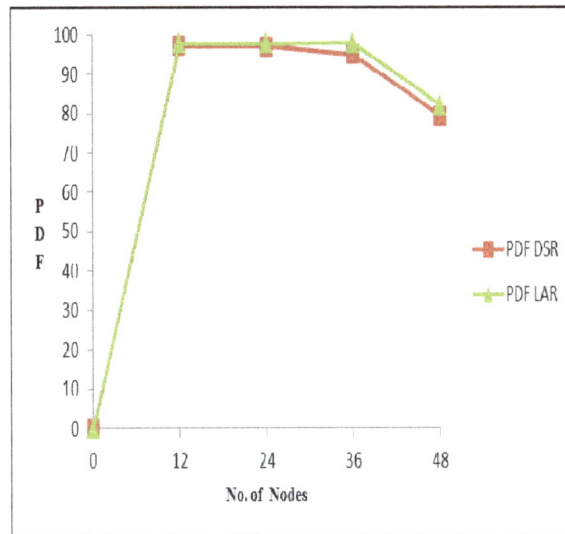

Figure.7 (a) PDF Vs Number of nodes

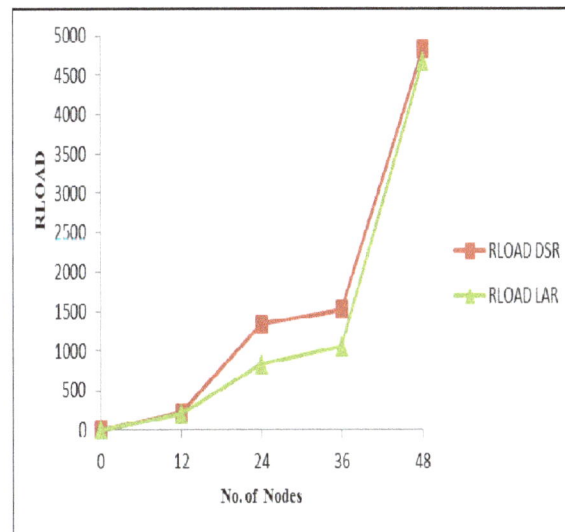

Fig.7 (b) RLOAD Vs Number of nodes

Figure.7 PDF and Routing Overhead (RLOAD) Vs Number of Nodes

4.3 . Under different fixed node mobility in MANET:

In a dynamic Ad-Hoc network the mobility of nodes can not be ignored. It has a vital role in MANETs. Mobile nodes in the simulation travel according to the random-way point mobility model. Each simulation runs for 100 seconds. Simulation is run with movement patterns generated for 0, 20,40,60,80 and 100 pause times. NS-2 simulation is done at 5- 20m/s mobility speeds with same 12 numbers of sources in the networks. A result for the same is given in following figures of this section.

4.3.1 Results:

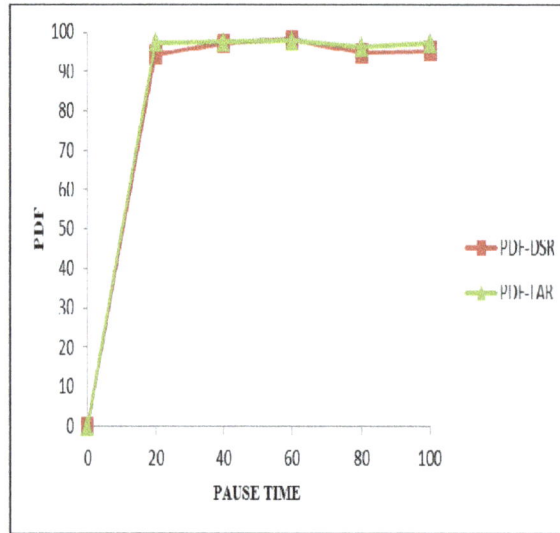

Figure.8 (a) PDF Vs Pause Time

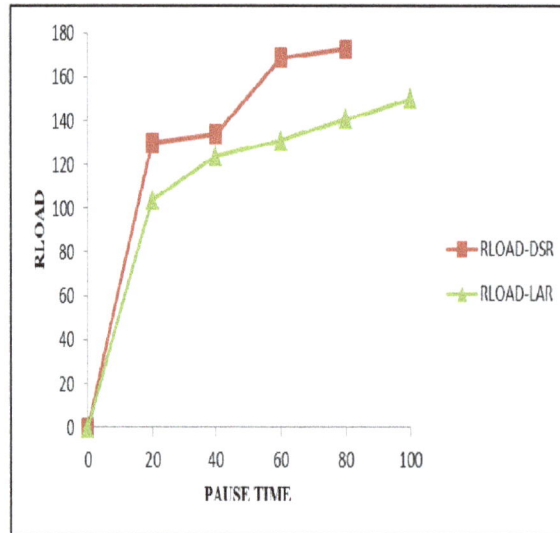

Fig.8 (b) RLOAD Vs Pause Time

Fig.8 PDF and RLOAD variations (5m/s)

4.3.2 Results:

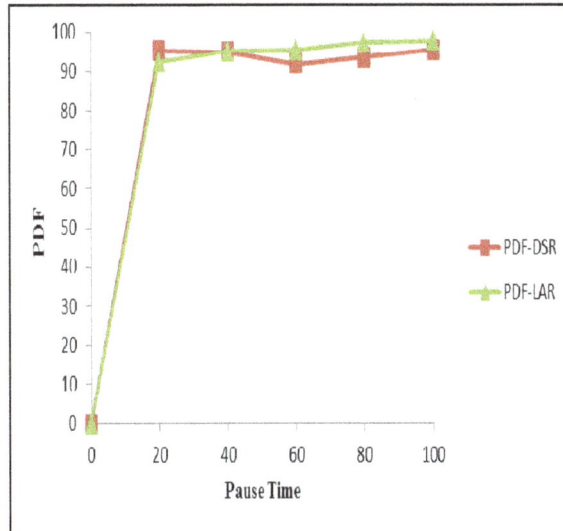

Fig.9 (a) PDF Vs Pause Time

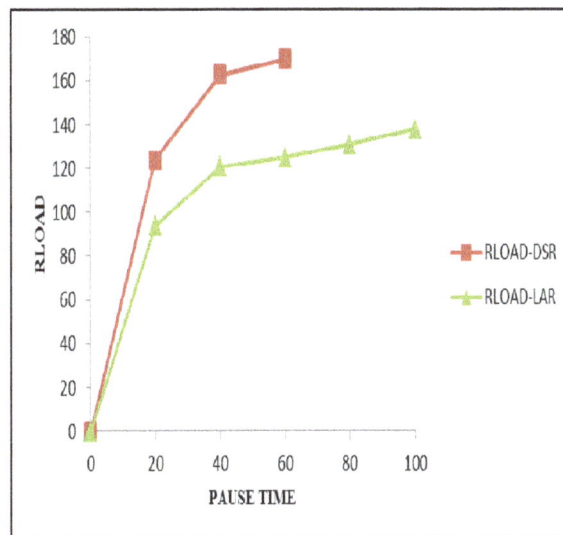

Fig.9 (b) RLOAD Vs Pause Time

Fig.9 PDF and RLOAD variations (10 m/s)

4.3.3 Results:

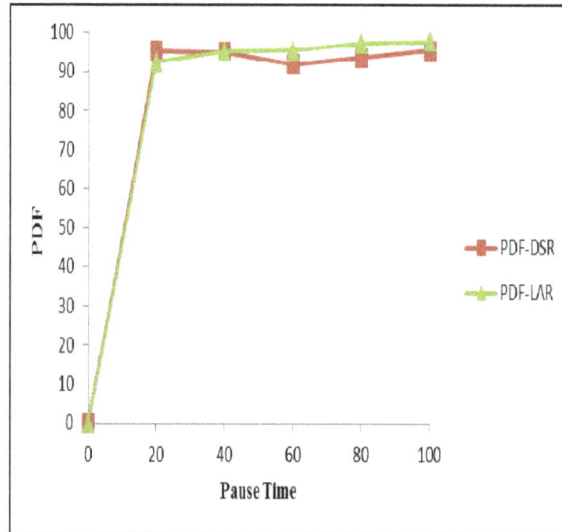

Fig. 10 (a) PDF Vs Pause Time

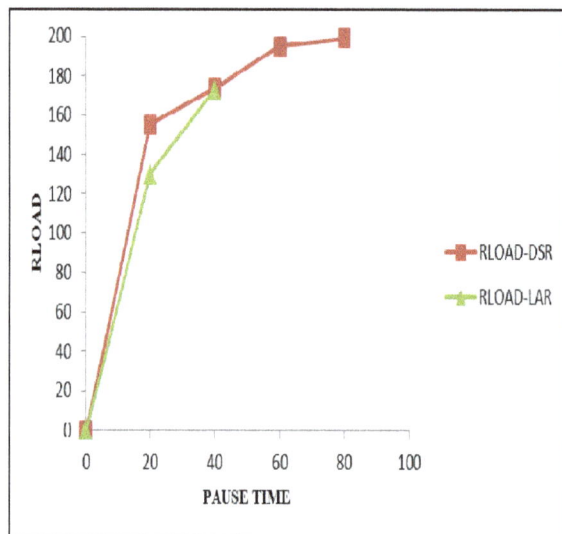

Fig.10 (b) RLOAD Vs Pause Time

Fig.10 PDF and RLOAD variations (15m/s)

4.3.4 Results:

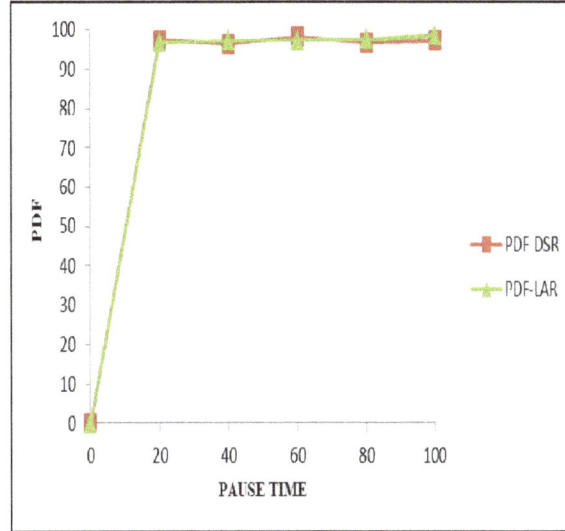

Fig. 11 (a) PDF Vs Pause Time

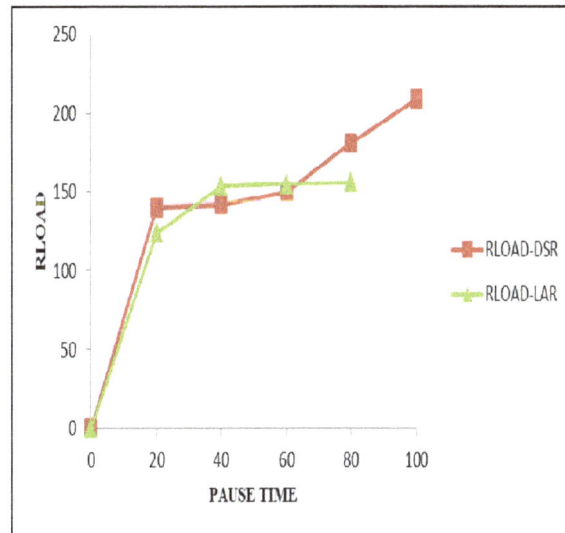

Fig.11 (b) RLOAD Vs Pause Time

Fig.11 PDF and RLOAD variations (20m/s)

From the above results, it is found that performance metrics are improved in a better ratio in high density networks at varrying mobility, that is the main advantage of position-based LAR protocol. It has been concluded from the evaluted results that performace metrics of MANET are better in case of LAR compared to DSR whenever number of nodes increases in the networks that is basic aim of this paper.

5. CONCLUSION

Ad-Hoc networks are characterized by multi-hop wireless connectivity and frequently changing network topology. There is a need for efficient dynamic routing protocols for MANETs; due to the scalability of the routing approach is an extremely essential. One of the approaches to scale up Ad-Hoc routing is geographical location-based routing (like a LAR). Because LAR floods only in the direction of expected destination istead of entire Ad-Hoc network. Hence routing overhead during LAR route discovery is to be minimized in other directions (except the direction of destination) and performance will be better in network. It is concluded that evalution of LAR protocol produces better results compared to DSR protocol in high density Ad-Hoc networks. Furthermore evalution of both was accomplished at varrying node mobility also in the same simultion scenerio. Finally it is concluded that LAR creats optimization of performance metrics such as, PDF, E2E Delay,routing overhead, NRL and lost data packets in MANETs. These optimization of parameters are important in case of Ad-Hoc network applications.

REFERENCES

[1]   Broustis, I., Jakllari, G. Repantis, T.; Molle, M. "A Comprehensive Comparison of Routing Protocols for Large-Scale Wireless MANETs", 3rd Annual IEEE Communications Society on Sensor and Ad Hoc Communications and Networks (SECON '06), Vol.3, pp 951–956, Sept. 2006.

[2]   Camp T., Boleng J., Williams B., Wilcox L., Navidi W., "Performance Comparison of Two Location Based   Routing Protocols for Ad Hoc Networks", vol.3, pp 1678-1687, INFOCOM 2002.

[3]   Das S.,Perkins C. and Royer E. "Performance Comparison of Two On-demand routing Protocols for Ad-Hoc network" Nineteenth Annual Joint Conference of the IEEE Computer and Communications Societies, IEEE INFOCOM, vol. 1,pp. 3-12, Tel Aviv, 2000.

[4]   David Oliver Jörg "Performance Comparison of MANET Routing Protocols in Different Network Sizes" Computer Science Project, Institute of Computer Science and Applied Mathematics Computer Networks and Distributed Systems (RVS) University of Berne, Switzerland 2003.

[5]   Dearham N., Quazi T. and McDonald S. "A Comparative Assessment of Ad-Hoc Routing Protocols" Proceedings of the South African Telecommunications Networks and Applications Conference (SATNAC), 2003.

[6]   E. Ahvar, and M. Fathy, "Performance Evaluation of Routing Protocols for High Density Ad-Hoc Networks based on Energy Consumption by GlomoSim Simulator" World Academy of Science, Engineering and Technology 29 2007.

[7]   Gupta N. and Gupta R., "Routing Protocols in Mobile Ad-Hoc Network: An Overview" Proc. International Conference on Emerging Trends in Robotics and Communication Technologies (INTERACT), IEEE Explore pp. 173-177, 3-5 Dec., Chennai, 2010.

[8]   Ko Y.B and Vaidya, N.H. "Location-Aided Routing (LAR) in Mobile Ad-Hoc Networks," Proc. IEEE MobiCom, Oct. 1998.

[9]   Mauve M. Widmer A. and Hartenstein H. "A Survey on Position-Based Routing in Mobile Ad-Hoc Networks" IEEE Network, vol. 15, Issue 6, pp.30-39, Nov.2001.

[10]   NS-2 Network simulator http://www.isi.edu/nsnam/ns.

[11]   Qabajeh L. Mat Kiah L. Qabajeh M., "A Qualitative Comparison of Position-Based Routing Protocols for Ad-Hoc Networks" IJCSNS International Journal of Computer Science and Network Security, VOL.9 No.2, pp. 131-140, February 2009.

12

A NOVEL ARCHITECTURE FOR SDN-BASED CELLULAR NETWORK

Md. Humayun Kabir

Department of Computer Science & Engineering,University of Rajshahi, Bangladesh.

ABSTRACT

In this paper, we propose a novel SDN-based cellular network architecture that will be able to utilize the opportunities of centralized administration of today's emerging mobile network. Our proposed architecture would not depend on a single controller, rather it divides the whole cellular area into clusters, and each cluster is controlled by a separate controller. A number of controller services are provided on top of each controller to manage all the major functionalities of the network and help to make the network programmable and more agile, and create opportunities for policy-driven supervision and more automation.

KEYWORDS

SDN, OpenFlow, LTE.

1. INTRODUCTION

Everyday new technology, policies and smart devices are emerging, todays networking concept is also developing accordingly. The traditional network infrastructure is considered as a single system made by many physical elements, such as routers, switches, and firewalls on which the whole network controlling activities depend for communication and services. A single modification in any part of the network can increase the maintenance effort on the whole network, and sometimes it may cause a miscarriage of the total network. At present, most of the IT related people identify the traditional networking paradigm as very much static and think it require a lot of effort to physically change and laboriously organize and legalize the network [1].

Software Defined Networking (SDN) is a new approach in the networking paradigm that has given the idea to deal efficiently with the emerging network and to better handle the major growth in data traffic, network virtualization, and mobility of user equipment [2] [3]. SDN generally permits network administrators/operators to regulate their network systems programmatically, serving them to improve capabilities and scale without compromising performance, reliability, or user experience [4].

The importance of networking is increasing day-by-day due to the emergent human's need and as a result it has been the key concept in the modern communication system. Now people are very much dependent on advanced technologies and innovative devices that are usually work through various communicating networks. Today's network provides all types communicating services and acts as the common information gateway to the whole world by sending and delivering messages, audios, videos, images and so on. A traditional network layout (shown in Figure 1) as it compares to an SDN network layout (shown in Figure 2) [5] is described in the following.

Traditional networking devices are composed of an embedded control plane that manages switching, routing and traffic engineering activities while the data plane forwards packet/frames based on traffic [6]. Here control plane is responsible to control the traffic related activities and data plane works as the traffic carrier. The control plane provides information used to build a forwarding table. The data plane consults the forwarding table to make a decision on where to send frames or packets entering the device. The networking device contains both of these planes and these are usually placed as built-in on the device [7].

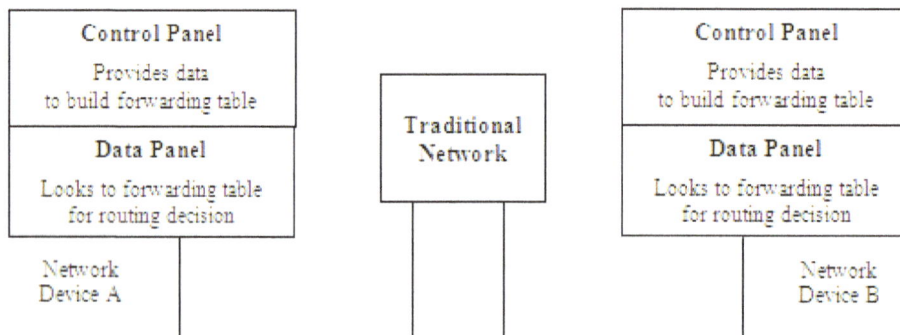

Figure 1. Traditional network layout

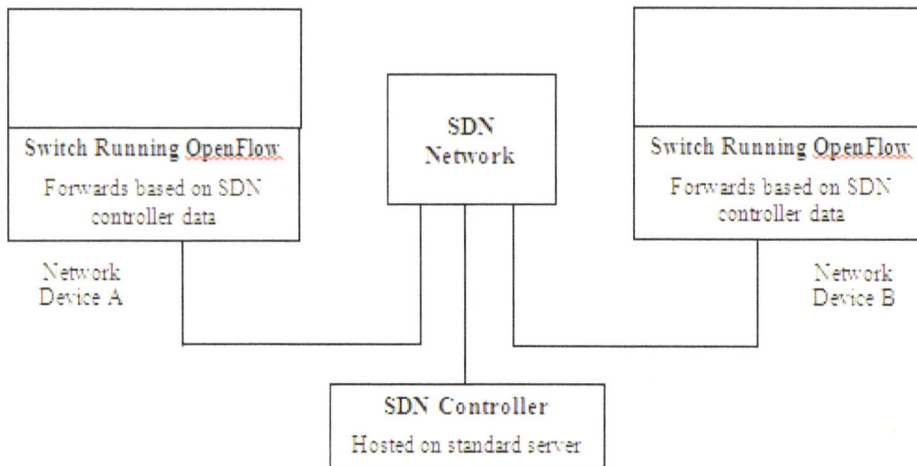

Figure 2. SDN network layout

In SDN architecture, control plane functions are removed from individual networking devices and hosted on a centralized server [8]. The SDN controller usually is an operating system with necessary SDN software. The controller generally communicates with the switch data plane through a protocol that is publicly known as OpenFlow [9]. OpenFlow transmits the instructions and commands to the data plane so that the data plane can forward the data to the right direction. To support the services the network devices must contain and run the OpenFlow protocol.

Mobile and wireless networks are growing rapidly and the technology behind them is changing continuously. As wireless devices become the main or even only option for more and more people to communicate with others, mobile operators must carry much volumes of traffic and at the same time provide a number of facilities or services. New cellular technologies, like Long Term Evolution (LTE) [10], have supported cellular providers/operators to maintain the stability of traffic growth by increasing the radio access volume. However, they now face a number of challenges of keeping up with the increasing demand in their core networks, which carry the User Equipment (UE) traffic between the Base Station (BS) and the Internet and the increasing number of wireless technologies in use simultaneously. Typical devices today support 3G and 4G cellular services as well as Wi-Fi and Bluetooth connectivity. To support these various types of services mobile operators usually have to manage increasing costs and handle operational headaches. In addition, carriers need flexible deployment choices to migrate from older to newer technologies without hampering the customer services.

The cellular and mobile network industry has been fighting to handle the growing data demands of new devices like smartphones and tablets from a number of years [11]. Future cellular networks are faced with the challenge of coping with significant traffic growth without increasing operating costs. SDN is a new networking approach that separates the control and forwarding planes of a networking device in a network [10-12]. This functional separation and the implementation of control plane functions on separate centralized platforms have been of much research interest due to various expected operational benefits [13].

In this article we propose a novel clustering SDN-based cellular network architecture that does not only depend on a single controller, rather it divides the whole cellular area into clusters, and each cluster is controlled by a separate controller. A number of applications or services are kept available on top of the controller that maintains all the controlling functions of the network. The controllers communicate and share information between them through a controller service. Basically, a controlling function is dependent on a number of services. In this way, much of the traffic and single-controller overwhelming could be minimized. To our knowledge, this will be the first work for cellular network that would utilize controller services efficiently by sharing their information rather than depending on only a central controller. The rest of the paper is organized as follows: section 2 briefly describes about the architecture of a generic cellular system, an overview of today's LTE/EPC cellular network architecture is demonstrated in section 3 and the ONF SDN reference model architecture is described in section 4. Related work and background study have been discussed in section 5. We have described our proposed architecture in section 6 and finally section 7 concludes our proposal.

2. CELLULAR NETWORK ARCHITECTURE

The architecture of a generic cellular system [14] is described in Figure 3. The schematic provides an idea of the different components in the traditional mobile network. The radio access subsystem is responsible to locate the position of the mobile station (MS). Sometimes these MSs are also called user equipments (UEs). Base stations (BSs – also called eNodeBs) are fixed transmitters that are points of access to the rest of the network. A MS keeps communication with a BS by sending and receiving information during idle period, cellular phone calls or other data transmission. Base stations are controlled by radio network controllers (RNCs) that are also responsible to manage the radio resources of each BS and MS (frequency channels, time slots, spread spectrum codes, transmit powers, and so on).

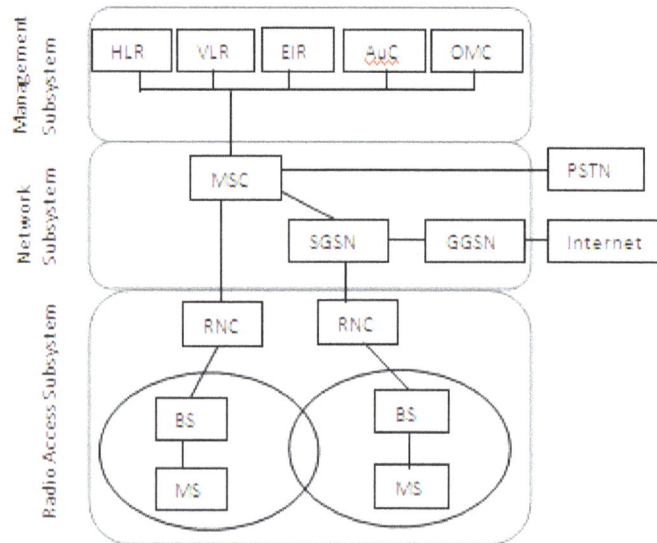

Figure 3. Generic cellular network architecture

The network subsystem is liable to carry voice and data traffic and also handles routing information of voice calls and data packets. The mobile switching center (MSC) and the serving and gateway GPRS (General Packet Radio Service) support nodes (SGSN and GGSNs) are responsible for handling voice and data respectively. These network entities control the mobility management; locate the cell or group of cells where a MS is positioned and update routing information when a MS makes a handoff. They connect to the public switched telephone network (PSTN) or the Internet. Several databases in the management subsystem are used for keeping track of the entities in the network that are currently serving the MS, security issues, accounting and other operations as shown in the upper part of Figure 3.

3. TODAY'S LTE CELLULAR DATA NETWORKS

In Long Term Evolution (LTE) cellular networks, a base station (eNodeB) generally connects to the Internet using an IP networking equipment [15], as shown in Figure 4. The user equipment (UE) directly makes a connection to a base station, which forwards traffic information through a serving gateway (S-GW) over a GPRS Tunneling Protocol (GTP) tunnel. The S-GW acts as a local mobility anchor point that maintains smooth communication when the user travels from one base station to another. The S-GW stores a large amount of state since users retain their IP addresses when they move from one location to another. The S-GW forwards traffic to the packet data network gateway (P-GW). The P-GW enforces quality of service policies and monitors traffic to perform billing. The P-GW also handles the connections to the Internet and other cellular data networks, and works as a firewall that blocks annoying traffic flow. The P-GW can handle different types of policies based on whether the user is travelling, features of the user equipment, usage caps in the service agreement, parental controls, and so on.

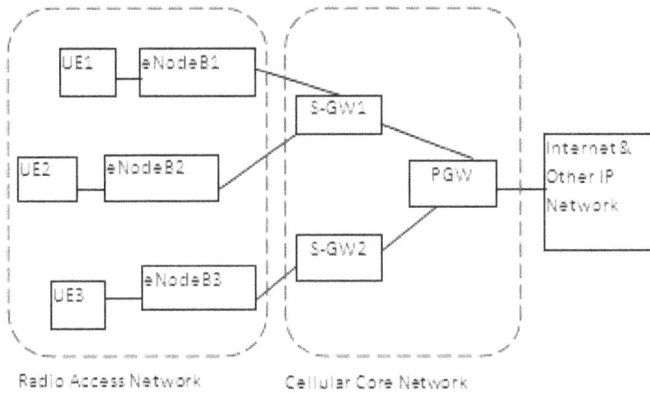

Figure 4. LTE data plane

Besides data-plane functionalities, the base stations, serving gateways, and packet gateways also join in several control-plane protocols, as illustrated in Figure 5. In coordination with the mobility management entity (MME), they handle hop-by-hop signaling to manage session setup, tear-down, and reconfiguration, as well as mobility e.g., location update, paging, and handoff. For example, in reply to a UE's request for dedicated session setup (e.g., for VoIP call), the P-GW forwards QoS and other session information (e.g., the TCP/IP 5-tuple) to the S-GW. The S-GW in turn sends the messages to the MME. The MME then requests the base station to assign radio resources and form the connection to the UE. During handoff of a UE, the source base station directs the handoff request to the target base station. After reception of an acknowledgement, the source base station transfers the UE state (e.g., buffered packets) to the target base station. The target base station also updates the MME that the UE has made new cells, and the previous base station to discharge resources (e.g., eliminate the GTP tunnel).

The S-GW and P-GW are also involved in routing policies by running protocols such as open shortest path first (OSPF). The Policy Control and Charging Function (PCRF) handle flow-based charging rules in the P-GW. The PCRF also offers the QoS authorization (QoS class identifier and bit rates) that chooses how to contact every traffic flow, based on the user's payment options. QoS policies and services can be dynamic, e.g. based on time of day. This must be imposed at the P-GW. The Home Subscriber Server (HSS) holds subscription data for each user, such as the QoS profile, any access constrains for roaming, and the associated MME. In the time of cell overloading, a base station cuts the highest rate allowed for subscribers according to their profiles, in coordination with the P-GW.

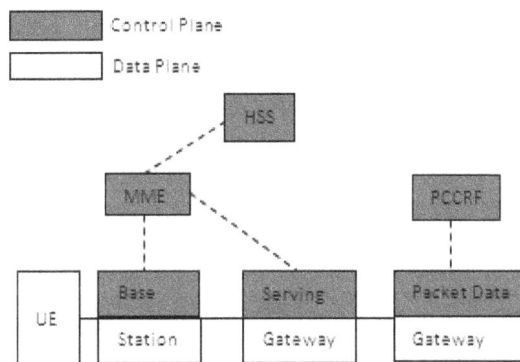

Figure 5. Simplified LTE network architecture

As today's cellular networks provide a number of services but their architectures have numerous major limitations. Centralizing monitoring activities, access control mechanisms, and quality-of-service policies at the packet gateway presents scalability challenges. This makes the networking devices or equipment very expensive (e.g., to purchase a Cisco packet gateway it usually requires more than 6 million dollars). Concentrating data plane activities at the cellular-Internet frontier forces all traffic related data through the P-GW, containing traffic between users on the same cellular network coverage, making it tough to host popular contents inside the cellular network. In addition, the network devices have vendor-specific configuration interfaces, and make communication through complex control-plane protocols, with a huge and increasing number of parameters under more restrictions (e.g., several thousand parameters for base stations). As such, network administrators or operators have limited control over the operation of their networks, with little ability to create innovative policies as well as to provide up-to-date services.

4. SDN OVERVIEW

Software Defined Networking is an innovative architectural approach in the networking arena that has been designed to allow more agile and cost-effective networks to provide network users the recent and future services. The Open Networking Foundation (ONF) is on the top position in SDN standardization, and has defined an SDN architecture model [16] as illustrated in Figure 6.

The ONF/SDN architecture model is comprised of three separate layers that are reachable through a number open APIs:

- The application layer consists of the end-user business applications that provide different communications services. Communications between the application layer and the control layer is managed by the API.
- The control layer controls and supervises the network forwarding functionality through an open interface.
- The physical layer usually contains the physical network devices or components (i.e. router, switch, etc.) that are responsible to handle packet switching and forwarding.

According to this architectural approach, the model is characterized and described by three key features:

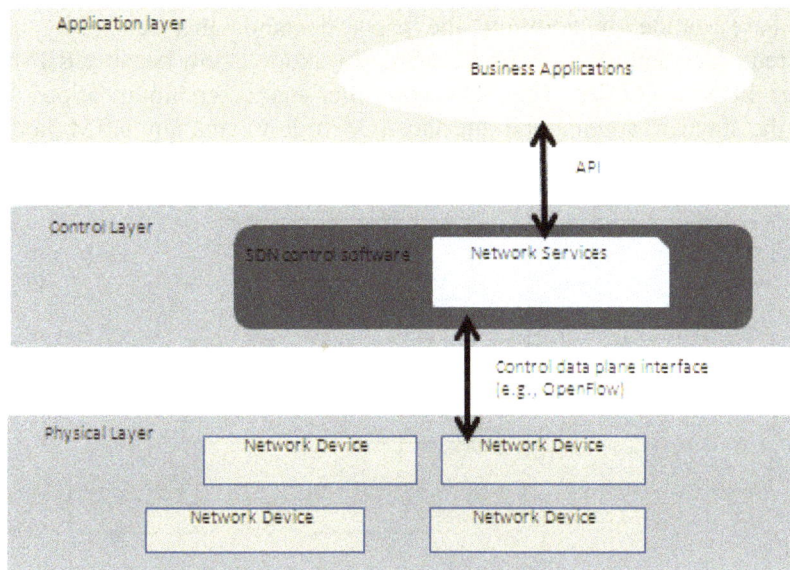

Figure 6. ONF SDN reference model

1. Logically centralized intelligence

   An SDN provides the full network overview from a single point of supervision using the standard interface OpenFlow [17]. By centralizing network functionalities or intelligence, all types of decision-making are performed based on a global (or domain) view where nodes are ignorant of the overall state of the network.

2. Programmability

   An SDN offers programmatic interfaces through different services that can automate and form network fabric configuration. SDN networks can attain revolution and variation from traditional networks by providing open APIs for applications to communicate and interconnect with the networks.

3. Abstraction

   In an SDN network, the business applications and services are abstracted from the underlying network technologies and mechanisms. Network devices are also abstracted from the SDN control layer to support portability for any application or services from any vendor or manufacturer.

5. RELTED WORK AND BACKGROUND STUDY

The ONF has defined an SDN architecture model for cellular network [18]. An SDN provides the overall network functionalities from a single point of administration using the standard interface, OpenFlow. ONF describes two use cases to illustrate the benefit of OpenFlow-based SDN for mobile networks:

- Inter-cell interference management
- Mobile traffic management

As shown in Figure 7, the logically centralized control layer provides radio resource allocation choices to be performed with global visibility through many base stations, which are more efficient than the distributed radio resource management (RRM), mobility management, and routing applications/protocols in use today. By centralizing network intelligence into the SDN controller, RRM decisions can be made based on the dynamic power and subcarrier allocation profile of each base station. In addition, the paper demands that scalability challenges are improved as the required compute capacity at each base station is low because RRM processing is centralized in the SDN controller. The SDN controller makes communications with the base stations through the standard southbound interface (OpenFlow), and any RRM modifications can be accomplished freely from the base station hardware.

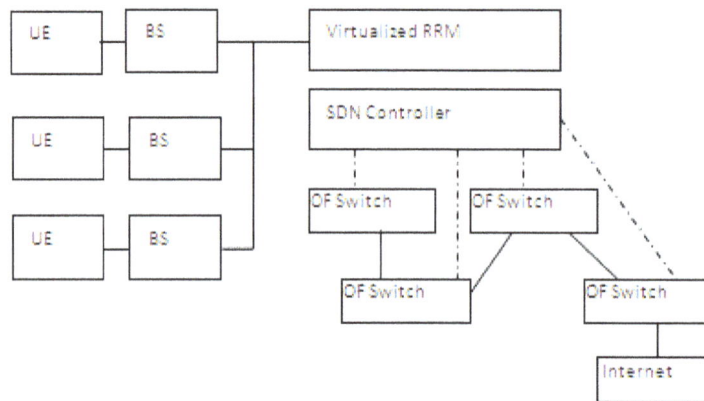

Figure 7. OpenFlow-enabled centralized base station control for interference management

Offloading is the term used in the networking that means moving traffic from a mobile network (cellular, small cells, femtocell) to a Wi-Fi network. It is also known as Wi-Fi roaming. The handover process is the power of software that enables networks with no loss of data/connectivity, preservation of IP address, etc. to maintain the user experience (UX). Offloading can also be applied in the reverse order. The OpenFlow controller (OF controller) will have to communicate with entities such as the ANDSF (access network discovery and selection function) for finding wireless networks close to the mobile user and performing the Wi-Fi offload (Figure 8). The destination selection of the roaming can be on the basis of a QoS metric such as performance, signal strength, or distance in order to maintain the UX.

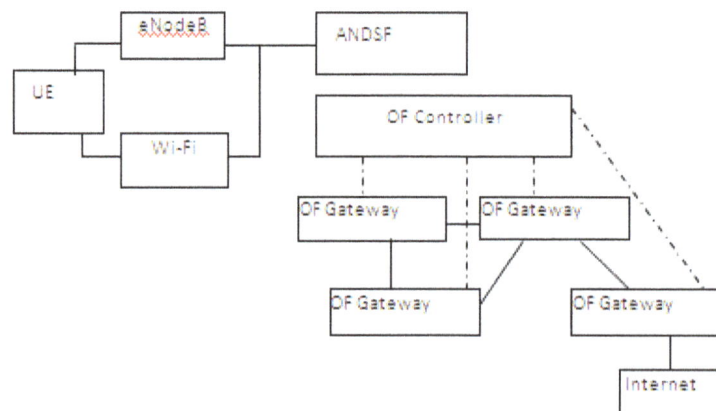

Figure 8. Openflow-based mobile offload

Cellular networks need an SDN architectural mechanism that provides fine-grain, real-time control without losing scalability. The authors in [15] propose four main extensions to SDN as shown in Figure 9, leading to the architecture for cellular network. They uses local controller with switch that communicate with the central controller. The main limitation of this approach is the management of the local and central controller as both may have dissimilar controlling information or data at the same time for a specific switch to forward packets.

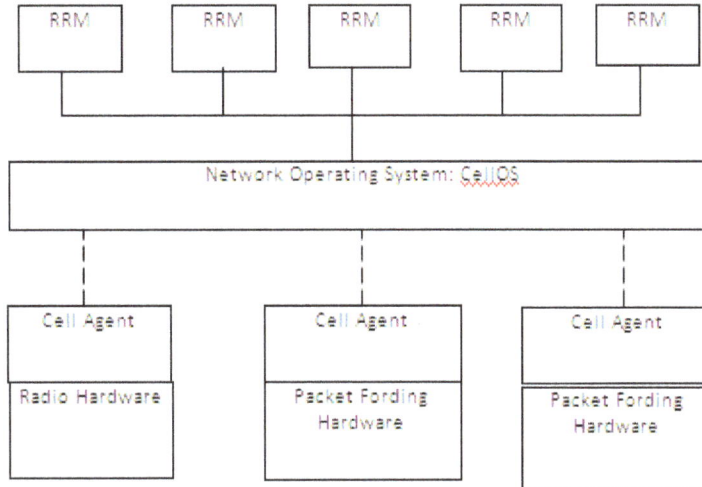

Figure 9. Cellular SDN architecture

SoftCell [19] is an SDN-based cellular network architectural model that demands to support a number of fine-grained services in a scalable manner for cellular core networks (Figure 10). In this article, the authors used local agents and access switch to each base station to communicate with the controller, they also used OpenFlow switches in the core network rather than EPC/LTE switches. It would be very difficult to deploy new software switches to each base station; also it may suffer same limitation as the above approach.

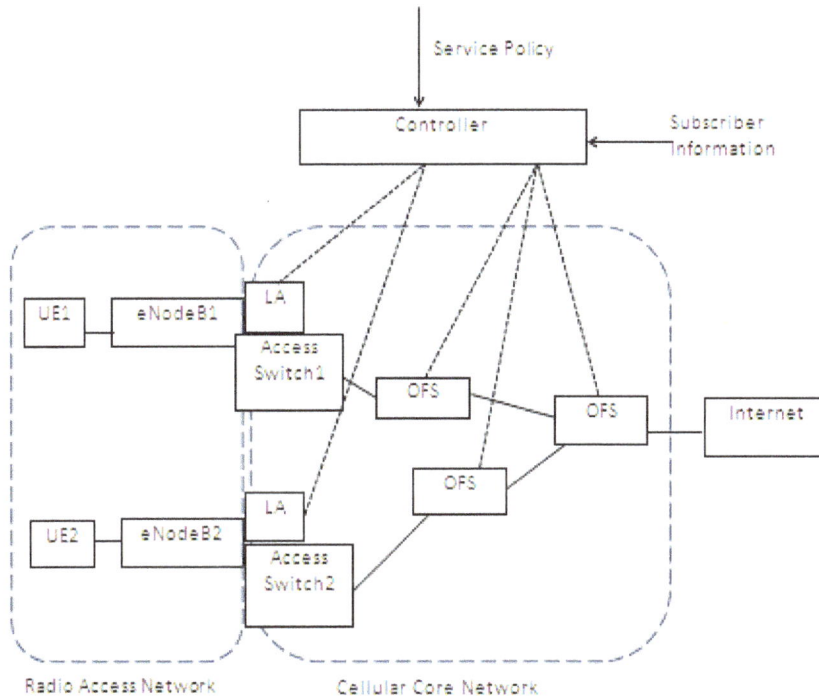

Figure 10. SoftCell Architecture

SoftRAN [20] is a SDN based centralized control plane architecture for radio access networks that localizes all base stations in a particular geographical area as a virtual big-base station comprised of a central controller and radio elements (individual physical base stations), but it does not apply any technique for cellular core network as depicted in Figure 11.

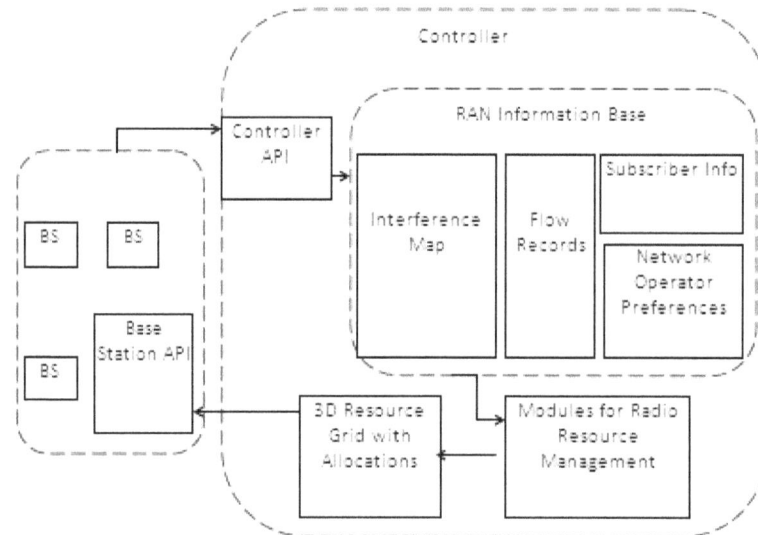

Figure 11. SoftRAN Architecture

In [21], the authors propose an SDN-based mobile networking approach integrated with legacy mobility control plane. They simply call this the partially-separated mobile SDN architecture that is compared to the fully-separated mobile SDN architecture where all the control is dominated by a SDN controller without taking the legacy mobility control plane into consideration (Figure 12). This paper is only for controlling the mobility of the user equipment's and they propose mobility control plane to each switch as like as local controller, and the central mobility controller acts as central controller same in the above techniques.

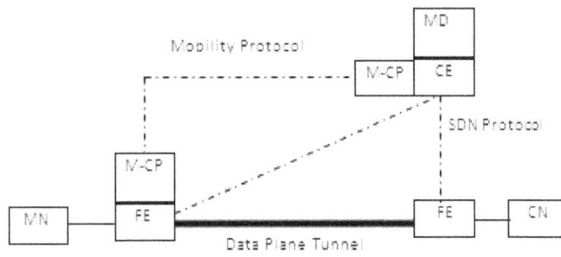

Figure- (a) Partially-distributed

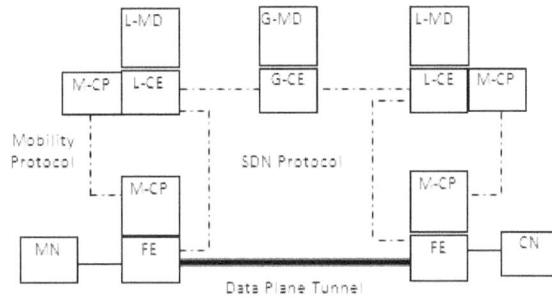

Figure- (b) Fully-distributed

Figure 12. SDN-based mobility management architecture

In the article [22], the authors present a new dynamic tunnel switching technique for SDN-based cellular core networks. This approach is to maximally utilize cloud and implement a virtualized EPC (Evolved Packet Core) serving and packet data network gateway (S/PGW) where control and user plane functions are separated from each other (Figure 13). They demand it would support 5G cellular network. The limitation is that it would not be able to adopt with today's existing network as well as future cellular network.

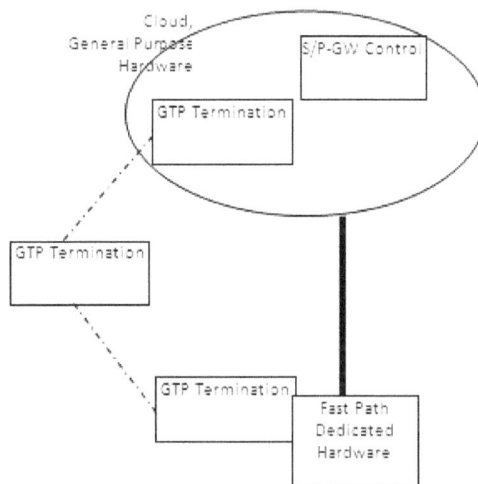

Figure 13. SDN-based virtualized S/P-GW

A new architecture for 3GPP LTE/EPC cellular network for on-demand connectivity service has been proposed in [23]. Their proposal also depends on a single SDN-based controller and used the expensive PGW switch to connect to the internet.

All the above proposals for SDN-based cellular network used only a single controller. As cellular data traffic has exploded in recent years and the rate will also be kept in the coming future, we think it would not be possible for a single controller to handle all the functionalities to manage the network. And hence, we have decided to establish a novel architecture for SDN-based cellular network that would be depended on a number of cluster-based controllers rather than a single controller.

6. PROPOSED ARCHITECTURE

To support the services of today's emerging cellular network and at the same time for future network, we propose a novel architecture for SDN-based cellular network. We change the control protocols on the interfaces of S1 (between MME and eNodeB), S11 (between MME and SGW) and S5 (between SGW-C and PGW-C) of the LTE/EPC architecture by the OpenFlow protocol. The other two interfaces S1 (from eNodeB to SGW-D) and S5 (from SGW-D to PGW-D) are controlled by the existing 3GPP protocol of the LTE/EPC architecture. According to the SDN principle we propose to separate the all controlling activities and place these to the central controller. The central controller is responsible to manage all the controlling functions through its different services that run on top of the OpenFlow controller.

Our proposed architecture needs not any change to the radio hardware at the base station; also it does not want extra support to connect to the Internet. The architecture comprises of the following entities as shown in Figure 14:

Controller: The SDN controller usually involves a network operating system (NOS) that handles a collection of application modules/services. The handling of a single packet may depend on multiple services. The NOS should be able to support the composition of the outcomes of multiple modules into a single set of packet-handling policies in each switch.

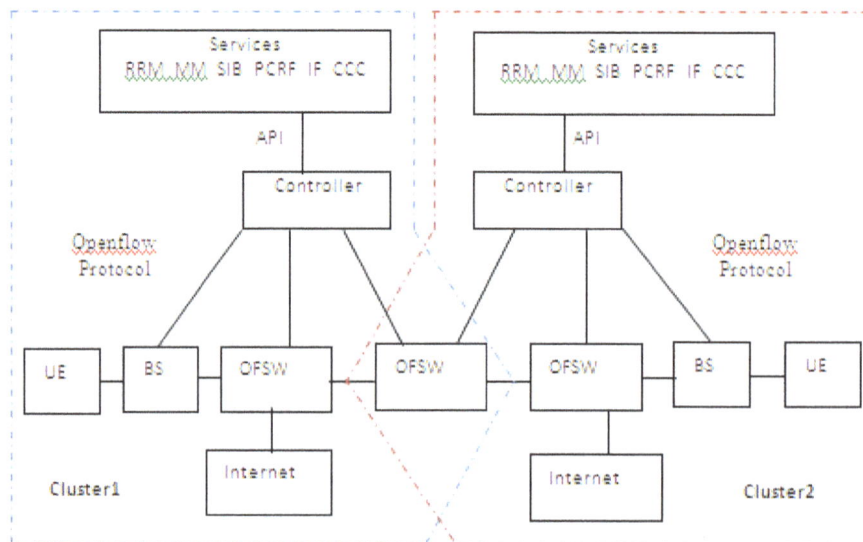

Figure 14. Proposed architecture

It is the key component of our architecture as it handles all controlling activities related to radio access network (RAN), the forwarding plane services of OpenFlow switches as well as gateways to the Internet. It implements high-level service policies by installing switch-level rules that direct traffic through middleboxes as in LTE/EPC network. The OpenFlow controller is responsible for user session establishment and load monitoring at the data plane. It controls all the controlling functions provided to the OpenFlow switches (OFSWs that resemble SGW-D and PGW-D in the LTE/EPC architecture) through different services that reside on top of it. In general, it interconnects with the services through the application programming interface (API) and holds the following modules:

- RRM: It maintains all the radio related functions provided to the base stations through the Openflow protocol rather than GTP tunneling protocol as used in the LTE/EPC architecture. Running radio resource management module (RRM) on top of a logically-centralized controller lets it much simpler to renovate in admission control, radio resource distribution, and interference management.
- MM: It is the service that provides UE authentication and authorization, and supports intra-3GPP mobility management (MM).
- SIB: The controller holds a subscriber information base (SIB) that stores and handles subscriber information, both for relatively static subscriber attributes and dynamic data of the user's current IP address, location, and total traffic consumption.
- IF: The infrastructure routing (IF) is responsible to identify the routes between pairs of network elements and the flow of traffic through the network depending on the subscriber's location (determined by the mobility manager).
- PCRF: The policy and charging rule function (PCRF) is responsible for traffic monitoring and packet scheduling activities.
- CCC: This module is the main part of controlling the clusters of our proposed architecture. When a cluster is heavy loaded, the controller generally sends packet to the switch that is shared between two clusters and search for that cluster which is less loaded by the controller-controller communication (CCC). All cluster that are neighbors, i.e. shared at least one switch, store these information in the CCC service table. Its main function is to update the controller status of the next cluster and share the information between controllers through the shared switch. And it is duty of the visiting cluster controller to forward packets which reached in the shared switch after updating its controller and CCC services. A packet only goes to the shared switch when the controller is extremely overloaded with huge traffic and it ensures only that the traffic is forwarded to another less crowded cluster. So by this way, many traffic as well as controller overwhelming is minimized.

OFSW: Today's OpenFlow switches (OFSW) already support many features needed in cellular networks [10]. These are commodity hardware switches that act like a SGW data plane (SGW-D) and are able to encapsulate/decapsulate GTP packets. This switch applies the rules received from the OpenFlow controller. It is responsible for packet forwarding between the eNodeB and SGW. Our proposed architecture also contains Openflow enabled a few gateway switches connected to the Internet. These gateway switches are much cheaper than PGW switches; they just perform packet forwarding, and relegate sophisticated controlling functions to the Openflow controller.

eNodeB: This is the base station that keeps the same radio functions specified by the 3GPP standard. It is enabled with the Openflow protocol for the data forwarding through the S1 (enodeB to SGW-D) interface. Therefore, the data forwarding is based on instructions received from the OpenFlow controller.

UE: These are mobile devices also called user equipments (UEs) that are today's LTE supported devices and work smart ways.

7. DISCUSSION AND CONCLUSION

Despite the extraordinary success of the cellular mobile telecommunications industry, many of the underlying design strategies and service assumptions that have served us arguably well over the past few decades may benefit from a fresh new look. Certainly, the LTE network architecture can eliminate a few network components, and simplifies some of the cellular network architectural compositions. Although, it was a change in the right direction, the result appears to provide somewhat constrained enhancements in terms of reduction in complexity and improvement in flexibility, as well as to maintain the heavy traffic of the today's popular cellular network.

An innovative SDN-based cellular network architecture has been proposed in this paper that does not only depend on a single controller, rather it splits the whole cellular area into a number of small clusters, and each cluster is managed by a distinct controller. A number of applications are provided on top of the controller that keeps all the controlling activities of the network. The controllers communicate and share information between them through a controller service. Basically, a controlling function is dependent on a number of services. As a result, much of the traffic and single-controller overloading could be minimized.

To design and meet the needs of the future mobile cellular network will be more difficult with a few general observations: there will be far smarter and new devices; the more base stations connecting them, and various numbers of applications - ever changing - running over the network. We believe that our proposed architecture would be able to fulfill the demands for today's and future cellular network and at the same time to support the challenges of this inevitability. Our proposed architecture would be simulated in the mininet emulator for future work to establish it as for practical or real world usage.

ACKNOWLEDGEMENT

The author thanks to the researchers whose research papers are included in this article.

REFERENCES

[1]   IBM Systems and Technology, "Software Defined Networking - A new paradigm for virtual, dynamic, flexible networking", October 2012
[2]   Kirk Bloede, "Software Defined Networking – Moving Towards Mainstream", Electronics Banking Research, August 2012
[3]   Brocade VCS Fabrics: The Foundation for Software-Defined Networks
[4]   "Network Transformation with Software-Defined Networking and Ethernet Fabrics", Brocade Communications Systems, Inc., 2012
[5]   Md. Humayun Kabir, "Software Defined Networking (SDN): A Revolution in Computer Network", IOSR Journal of Computer Engineering (IOSR-JCE), Volume 15, Issue 5 (Nov. - Dec. 2013), PP 103-106
[6]   "Software Defined Networking: What Is It and Why Do You Need It?", enterasys secure network
[7]   http://globalconfig.net/software-defined-networking-vs-traditional/
[8]   Open Networking Foundation, "Software-Defined Networking: The New Norm for Networks"
[9]   http://www.brocade.com/solutions-technology/technology/software-defined-networking/openflow.page
[10]   "LTE (telecommunication)", From Wikipedia, the free encyclopedia.
[11]   B. Kim & P. Henry, "Directions for Future Cellular Mobile Network Architecture", AT&T Labs - Research.

[12] Philip Bridge, "Revolutionizing Mobile Networks with SDN and NFV", Cambridge Wireless Virtual Networks SIG, 8th May, 2014.

[13] "NFV", http://www.etsi.org/technologiesclusters/technologies/nfv

[14] David Tipper, Prashant Krishnamurthy, and James Joshi, "Network architecture and protocols for mobile positioning in cellular wireless systems", Department of Information Science and Telecommunications, University of Pittsburgh, Pittsburgh, PA 15260.

[15] Li, L. E., Mao Z. M., Rexford J. "Toward Software-Defined Cellular Networks", In Proceedings of IEEE EWSDN. 2012.

[16] "ONF", https://www.opennetworking.org/

[17] "OpenFlow Switch Specification", version 1.3.2. Open Networking Foundation. 2013.

[18] "OpenFlow™-Enabled Mobile and Wireless Networks", ONF Solution Brief, September 30, 2013.

[19] Jin X., Li L. E., Vanbever L., Rexford J. SoftCell: Scalable and Flexible Cellular Core Network Architecture. In Proceedings of ACM CoNEXT. 2013.

[20] Aditya Gudipati, Daniel Perry, Li Erran Li, Sachin Katti, "SoftRAN: Software Defined Radio Access Network", HotSDN'13, August 16, 2013, Hong Kong, China.

[21] Seil Jeon, Carlos Guimarães, Rui L. Aguiar, "SDN-Based Mobile Networking for Cellular Operators", MobiArch'14, September 11, 2014, Maui, Hawaii, USA.

[22] Johanna Heinonen, Tapio Partti, Marko Kallio, Kari Lappalainen, Hannu Flinck, Jarmo Hillo, "Dynamic Tunnel Switching for SDN-Based Cellular Core Networks", AllThingsCellular'14, August 22, 2014, Chicago, IL, USA.

[23] S B H said, M R Sama, K Guillouard, L Suciu, "New control plane in 3GPP LTE/EPC Architecture for on-demand connectivity service", Proc. Of IEEE CloudNet 2013.

13

AN EFFICIENT MODEL FOR REDUCING SOFT BLOCKING PROBABILITY IN WIRELESS CELLULAR NETWORKS

Edem E. Williams[1] and Daniel E. Asuquo[2]

[1]Department of Mathematics/Statistics and Computer Science
University of Calabar, PMB 1115, Calabar-Nigeria

[2]Department of Computer Science
University of Uyo, PMB 1017, Uyo-Nigeria

ABSTRACT

One of the research challenges in cellular networks is the design of an efficient model that can reduce call blocking probability and improve the quality of service (QoS) provided to mobile users. Blocking occurs when a new call cannot be admitted into the network due to channel unavailability caused by limited capacity or when an ongoing call cannot be continued as it moves from one base station to another due to mobility of the user. The proposed model computes the steady state probability and resource occupancy distribution, traffic distribution, intra-cell and inter-cell interferences from mobile users. Previously proposed models are reviewed through which the present model is built for use in emerging wireless networks so as to obtain improved QoS performance. The developed model is validated through simulations in MATLAB and its equations implemented using Java Programming Language. The results obtained indicate reduced call blocking probability below threshold.

KEYWORDS

Call Blocking Probability Reduction, Cellular Networks, Quality of Service & Computer Simulations

1. INTRODUCTION

In cellular mobile networks, call blocking can occur as either new call blocking or handoff call blocking. The former refers to blocking of a new call request due to lack of available channel while the later refers to blocking of a call in the new cell as the mobile moves from its originating base station (BS) to a new BS. Ideally, during handoff or handover, the distributed mobile transceivers move from cell to cell during an ongoing continuous communication and switching from one cell frequency to a different cell frequency is done electronically without interruption and without a BS operator or manual switching. Typically, a new channel is automatically selected for the mobile unit on the new BS which will serve it. The mobile unit then automatically switches from the current channel to the new channel and communication continues. The two kinds of arriving calls to a cell site are shown in figure 1. Call blocking probability is one of the quality of service (QoS) parameters for performance evaluation in wireless cellular networks. According to [1], for better QoS it is desirable to reduce the call blocking probability. Call blocking is perceived negatively by users because it results to degradation in required QoS and developing an efficient model to reduce its probability in cellular networks is a growing research aimed at improving overall cellular system performance.

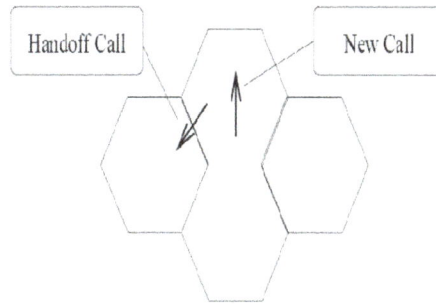

Figure 1: The two kinds of arriving calls to a cell site

The existing Visafone network suffers more of call blocking causing more calls to be rejected due to channel unavailability, as shown in figure 2. It indicates that only a few of its BSs within the region under study were able to carry traffic load beyond 400 Erlang per cell. This is definitely not a good characteristic of a reliable and efficient network. The key Performance Indicators (KPIs) usually measured by the regulatory bodies or commissions include Call Setup Success Rate (CSSR), Call Completion Rate (CCR), Standalone Dedicated Control Channel and Handover Success Rate (SDCCH), Call Data Rate (CDR) and Traffic Channel Congestion with or without Handover (TCHCon). The network operators could be rated on excellent, good, improvement, fluctuation, poor, slight decay and mostly steady below threshold on the KPIs. It is important to note that a rating of improvement does not mean the target key performance indicator is met. Rather, it means that the trend to reach the threshold is progressing towards the set target of the indicator, taking into consideration the challenges the operators are facing daily. It is therefore important that the operators continue working towards meeting the set target by adopting improved architectures despite challenges faced in infrastructure, upgrade, and service delivery.

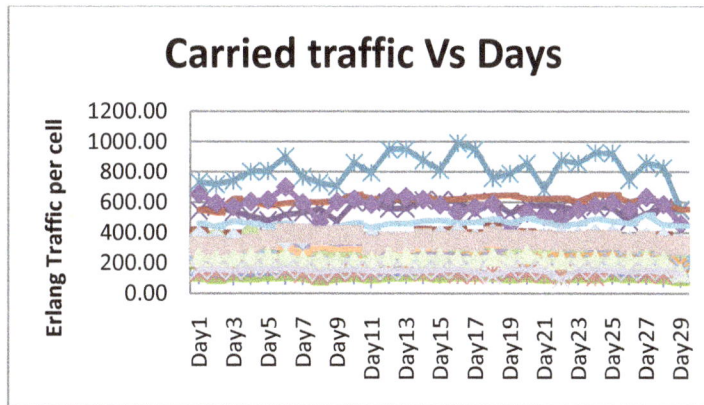

Figure 2: Visafone BS's Carried Traffic Load for 30 days

Several proposed models or techniques for reducing call blocking probability exist in literature either for reducing new call blocking [2], or reducing handoff blocking [3,4,5] or reducing both [6, 7, 8]. Whatever the case, there is a tradeoff between reducing handoff probability and new call blocking probability. A proposed method which attempts to optimally reduce handoff failures in mobile networks without significant increase in blocking probability of originating calls within a cell is highly desirable. However, call blocking model could either be said to be hard or soft. In hard call blocking models, the blocking probability is evaluated with fixed-valued parameters regardless of the channel and traffic conditions. In a Code Division Multiple Access (CDMA)-based network, the interference-limited nature makes it difficult to achieve accurate admission control. Because of the co-channel interference, the amount of resources (power, bandwidth) required by each user is dependent on the number of users in the system, their geographical locations, and physical channel conditions. In soft blocking models, the blocking probability is evaluated taking note of interference nature of the CDMA network. This work develops and implements a soft blocking probability reducing model for new call request in the uplink of a CDMA cellular network. A well established CDMA network named Visafone is studied in the South-South Zone of Nigeria's Niger Delta region.

The rest of this paper is organized as follows: section 2 reviews existing literature on call blocking probability estimation for both hard and soft blocking while section 3 presents the system design for the proposed model for reducing soft blocking probability with the developed algorithm. In section 4, the pseudo code for the developed algorithm is presented, the model's accuracy is validated through computer simulations in MATLAB and the equations implemented in Java. Section 5 presents the simulation results and conclusion giving direction for future works.

2. LITERATURE REVIEW

In a wireless network composed of several BSs serving some mobile users, user's power is limited to some given maximal value. The same frequency spectrum is available to all BSs (i.e., the frequency reuse factor is unity, 1). For a CDMA network, the interference of single user detection is regarded as noise. MSs and BSs are both assumed to be uniformly and randomly distributed on an infinite plane, but with different terminal densities (coverage areas). Blocking occurs in a network when due to limited capacity at least one link on the route is not able to admit a new call. Thus, such a user will not be able to subscribe to a particular channel. The following section discusses the two types of call blocking for CDMA BSs.

2.1 Hard Call Blocking

Hard blocking occurs when arriving calls to a network are blocked due to lack of available channels irrespective of the traffic characteristic and channel conditions [2, 5].

2.1.1 Erlang-B Model

The Erlang-B formula in equation (1) is used to compute hard blocking as a function of the number of available channels and the offered load. The equation is based on analytical probability theory and can be used when the following assumptions are satisfied.

    i.   All call attempts are Poisson distributed with exponential service time
    ii.   Blocked calls are cleared (BCC) in the system and that the caller tries again later

$$P_B = \frac{A^N/N!}{\sum_{i=0}^{N}\left(A^i/i!\right)} \tag{1}$$

In equation (2.3), P_B = probability of blocking,
    A = offered traffic in Erlang,
    N = number of channels in the cell, and
    i = number of busy channels
Thus, there is no queuing and no retry for unsuccessful calls with the Erlang-B model.

2.1.2 Extended Erlang B Model

The Extended Erlang B Model uses the same formula and assumptions as Erlang-B model except that a percentage of callers retry their calls until they are serviced. This model is commonly used for standalone trunk groups with a retry probability such as a MODEM pool. The model is as expressed in equation (2).

$$P_b = \frac{A^N/N!}{\sum_k^N A^k/k!} \qquad (2)$$

Where, k is the number of busy channels, N is the number of servers (trunks), A is the traffic density in Erlang, and P_b is the blocking probability.

The Extended Erlang B traffic model is used by telephone system designers to estimate the number of lines required for public switch telephone network connections (trunks) or private wire connections and takes into account the additional traffic load caused by blocked callers immediately trying to call again if their calls are blocked. It therefore, allows retry of unsuccessful calls and assumes infinite population of callers (sources). This traffic model may be used where no overflow facilities are available from the trunk group being designed.

2.1.3 Erlang-C Model

In the Erlang C Model, the system is designed around the queuing theory. The caller makes one call and is held in a queue until answered, so the formula expresses the waiting probability. Just as the Erlang B formula, Erlang C assumes an infinite population of sources, which jointly offer traffic of A Erlangs to N servers. However, if all the servers are busy when a request arrives from a source, the request is queued. An unlimited number of requests may be held in the queue in this way simultaneously.

This formula calculates the probability of queuing offered traffic, assuming that blocked calls stay in the system until they can be handled. This formula is used to determine the number of agents or customer service representatives needed to staff a call centre a specified desired probability of queuing. The blocking probability or the delay probability (waiting probability) is given in equation 3.

$$P_w - \frac{(N.\,A^N)/N!\,(N-A)}{\sum_{i=1}^{N-1} A^i/i! + (N.\,A^N)/N!(N-A)} \qquad (3)$$

where,
  A is the total traffic offered in units of erlangs
  N is the number of servers; i is the number of busy servers
  P_w is the probability that a customer has to wait for service

It is assumed that the call arrivals can be modeled by a Poisson process and that call holding times are described by a negative exponential distribution. A common use for Erlang C is modeling and dimensioning call center agents in a call center environment. It can also be used to determine bandwidth needs on data transmission circuits.

2.1.4 Engset Formula

The Engset formula is used to determine the probability of congestion occurring within a telephony circuit group. It deals with a finite population of S sources rather than the infinite population of sources that Erlang assumes. The formula requires that the user knows the expected peak traffic, the number of sources (callers) and the number of circuits in the network.

Engset's formula given in equation (4) is similar to the Erlang-B formula; however one major difference is that the Erlang's equation assumes an infinite source of calls, yielding a Poisson arrival process, while Engset specifies a finite number of callers. Thus Engset's equation should be used when the source population is small. But for population sources greater than 200 users, extensions or customers, it becomes similar to Erlang-B model [9, 10].

$$P_b = A^N \binom{S}{N} \Big/ \sum_{i=0}^{N} A^i \binom{S}{i} \tag{4}$$

where,
A = offered traffic intensity in Erlangs, from all sources
S = number of sources of traffic
N = number of circuits in group
Pb = probability of blocking or congestion

In the traditional flat cellular networks, the Erlang-B model is generally used to describe the limitation in physical resources independent of the quality experienced on the radio interface (i.e. hard blocking). With the increasing complexity of CDMA cellular networks, the required assumptions are no more valid and the Erlang-B formula is found to overestimate the capacity. Thus, to properly account for the quality of service experienced at the BS, soft blocking should be modeled and evaluated.

2.2 Soft Call Blocking

Soft blocking is related to the amount of interference in a network. There may be plenty of channels available at a BS but since there are many users in the same cell already, the interference level is such that adding an additional user would increase the interference above a predetermined threshold. The call is therefore denied. Modeling soft blocking due to interference is important and is considered a major aspect of this paper.

In [11], three explicit analytical models (single random trials (SRT), repeated random trials (RRT), and least busy fit (LBF)) were developed for evaluating the request blocking probability of movie files in video-on-demand (VoD) systems under three server selection schemes. The authors reported that the choice of server selection schemes can significantly affect the blocking probability performance of the system, and validated the accuracy of the analytical models through simulation.

A rapid and accurate method for evaluating the quality of service (QoS) perceived by the users in the uplink of wireless cellular networks was proposed in [2]. In doing so, the author aimed at accounting for the dynamics induced by the arrivals and the departures of users. The evaluated QoS was in terms of the blocking probability for streaming users and the throughput for elastic calls. The blocking probability of streaming users was evaluated using the Kaufman-Roberts algorithm as in [12, 13], whereas the throughput of elastic calls was evaluated using a multi-class processor sharing model. The research in [14] modeled soft blocking in multi-cell CDMA systems as an independent birth and death process at each cell. The model predicts the distribution of the number of calls connected to a base station.

In [15], the authors presented an analytical model for the estimation of the blocking probability as a function of the offered traffic per user in a cellular environment, where capacity is determined by hard blocking and the average number of users per cell is small. Using statistical model, they concluded that the number of mobiles audible to a base station with the strongest signal has a Poisson distribution, which mean is given in terms of the mean densities of mobile and base stations and the parameters of the attenuation law. The model does not represent a tool for detailed network planning.

In [16], the outage probability is considered a performance measure for real-time traffic in wireless networks. They observed that the blocking and outage probabilities do not have closed-form expressions as they strongly depend on the traffic characteristics (call duration, bit rate requirement, etc.), the radio conditions (fading, shadowing, noise, interference, etc.), the considered admission and outage policies. They assumed that the admission and outage policies satisfy a certain monotonicity property. Their results are applied to the uplink and the downlink of CDMA networks. In [17], a wireless network with beam-forming capabilities at the receiver is considered. They derived the blocking probabilities for calls in the system, under different traffic policies. For a set of co-channel transmitters, their success probabilities for being captured by separate antenna beams are computed. These success probabilities are taken into account in the queuing model of the system. Their analytical and numerical results show that adaptive beam-forming at the receiver reduces the blocking probability of calls and increases the total carried traffic of the system.

In [18], a new resource-dimensioning concept based on both the allowable noise-rise and traffic statistics is presented. The soft blocking probability based on outage probability and the assumption of the Poisson arrival and exponential services time are first derived, with a consistent view on traffic dimensioning. The relationship between outage probability, soft blocking probability and hard blocking probability is discussed. The authors in [19] focused on the call blocking probabilities calculation in a WCDMA cell with fixed number of channels and finite number of traffic sources. They proposed the use of the Engset Multi-rate Loss Model (EnMLM) in the uplink direction, which incorporates local blockings. The call admission depends on the availability of the required channels. To analyze the system, they formulated an aggregate one-dimensional Markov chain.

The work in [4] studied the QoS in terms of blocking and dropping probabilities, but the interference between the users was not taken into account explicitly. The work in [20] studied the QoS in wireless local area and sensor networks whereas this work focuses on wireless cellular networks. Certain models for soft blocking evaluation assumes that there is a constant number of users N in the cell, power control is perfect, and each user requires the same signal bit energy to noise spectral density E_b/N_o. From the above review, it appears studies on soft blocking have not considered interference under imperfect power control which this work considers by extending the Kaufman-Roberts algorithm under the assumption that in reality, none of these assumptions holds. The reason is that the number of active users in a cell is Poisson distributed with mean arrival and exponential service time λ/μ. Furthermore, due to voice activity, each user is ON with probability, v and OFF with probability $(1 - v)$ and each user requires a different E_b/N_o to achieve a desired bit error rate or communication signal quality.

3. PROPOSED MODEL FOR REDUCING CALL BLOCKING PROBABILITY

In CDMA systems, signals of each MS can be modeled as interfering noise for the others, leading to degradation in service. Adopting imperfect power control in CDMA wireless networks is to regulate the transmission power levels of MS such that each user obtains a satisfactory QoS. This goal is more precisely stated as to achieve a certain SINR regardless of channel conditions while minimizing the interference and battery usage, and hence improving the overall performance. Modeling the uplink system interference involves taking into consideration the cell model, path loss model, power control, and simulation parameters. The proposed system design is shown in figure 3 and implemented in six algorithmic stages. The stages are:

(i) modeling the state of a cell
(ii) Computation of total uplink interference
(iii) Computation and comparison of maximum effective cell load with threshold
(iv) Computation of soft blocking probability when cell load threshold is exceeded
(v) Blocking or accepting new call based on computed probability
(vi) Storage of generated results and updated system parameters

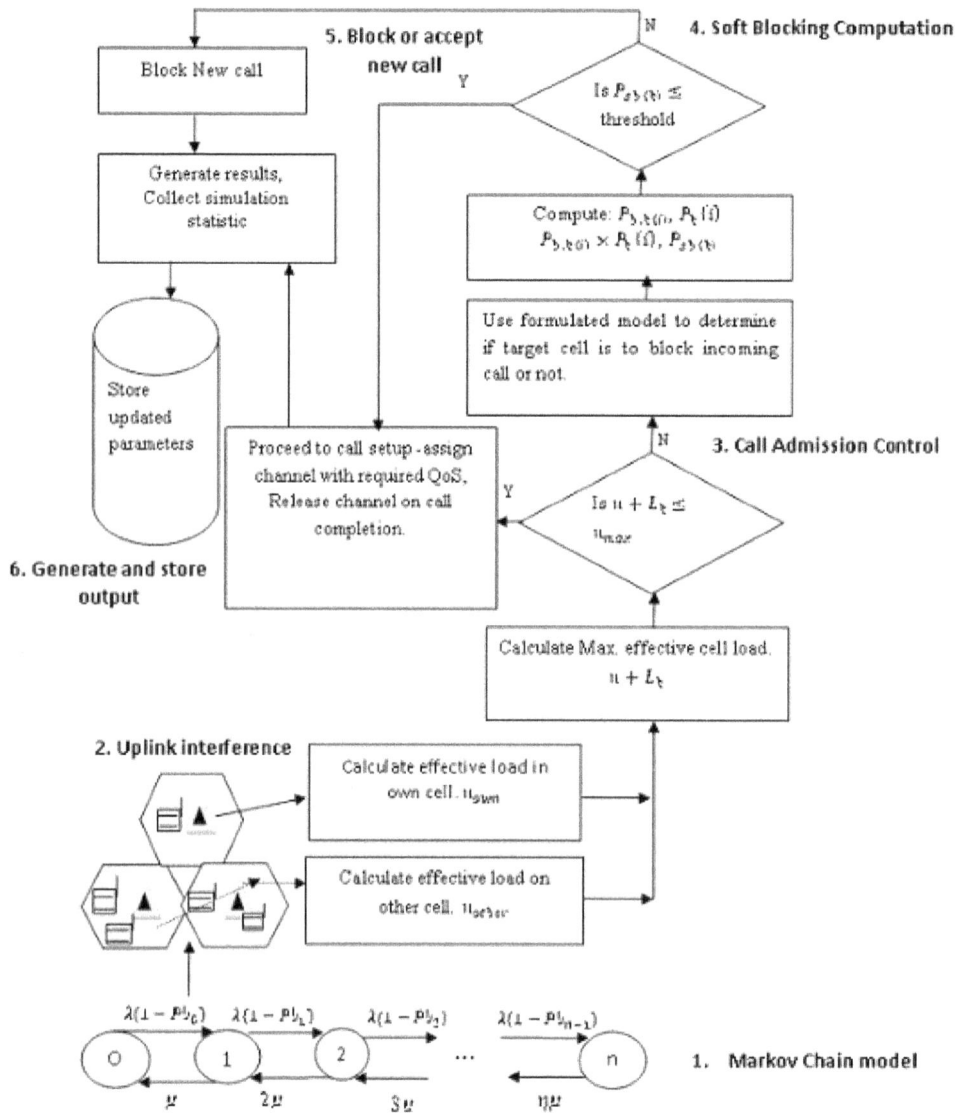

Figure 3: Proposed System Design for Soft Blocking Probability Evaluation

Stage1: Modeling the State of a Cell

Cell Model: In our model, each cell blocks newly arriving calls with a state dependent probability illustrated using Markov Chain in figure 3. The state of a cell is the number of users currently in that cell. Other cells simply contribute interference which causes blocking with some probability, $P_{b(i)}$ which is assumed to depend on the state i of the current cell. We model the state of a cell as a birth and death process and showed that the arrival rate is thinned by the blocking probability, $\lambda(1 - P_{b(i)})$, while the departure rate is iu.

Stage 2: Computation of Total Uplink Interference

CDMA systems consider soft blocking taking note of interference from mobiles in own cell and other cells. Therefore in the uplink, the interference experienced by a certain mobile is related to the load distribution within the network. The total interference is computed considering own-cell interference, other-cell interference, and interference due to an empty system referred to as thermal noise i.e $I_{total} = I_{own} + I_{other} + N_0$. This is represented in figure 3 in the hexagonal 3-cell structure. The more users are active at the BS, the larger is the multi-access interference at the BS and the higher are the transmit powers required by mobiles to fulfill their E_b/N_o requirements.

SINR Calculation: The signal-to-interference-noise ratio $SINR$ of each and every connection in the uplink depends on the power emitted by mobile users, own-cell interference I_{own}, other-cell interference I_{other}, thermal noise N_0, and multi user detection factor, β as given in equation 5.

$$SINR_{UL} = \frac{S}{(1-\beta)I_{own}+I_{other}+N_0} \qquad (5)$$

where the received signal, $S = P_k - \max(L - G_{tx} - G_{rx})$, and P_k is the transmit signal power, G_{tx} is the transmitting antenna gain, G_{rx} is the receiving antenna gain and L is the path loss propagation (attenuation) model from MS-BS.

Path Loss Model: Considering a BS antenna height Δh_b of 15metres and log-normally distributed shadowing (logF) with standard deviation of 10dB, the path loss is calculated based on equation (6) as in [21] as follows:

$$L = 40(1 - 4 \times 10^{-3}\Delta h_b)\log(d) - 18\log(\Delta h_b) + 21 \times \log(f) + 80 \qquad (6)$$
$$L = 127.8 + 37.6\log(d) + \log(F) = 137.8 + 37.6\log(d) \qquad (7)$$

where d is the MS-BS separation in kilometers.

Users connect to the BS per cell and each mobile connecting to one BS only at any given time. The i_{th} mobile transmits with a nonnegative uplink power level of $0 \le P_i \le P_{max} \, \forall_i$, where P_{max} is a sufficiently large upper-bound imposed for technical reasons. The received power at the l_{th} BS, x_{il}, is the attenuated version of the transmitted power level, $x_{il} = h_{il}P_i$, where the quantity h_{il} $(0 < h_{il} < 1)$ represents the slow-varying channel gain (excluding any fading).

Stage 3: Computation and Comparison of Maximum Effective Cell load with Threshold Power Model: When a call arrives to the cell, the noise rise is estimated and if it exceeds a maximum predefined threshold, the call is blocked and lost. Noise rise is the ratio of total received power at the BS, I_{total}, to the thermal noise power, N_o given in equation (8) as:

$$NR = \frac{I_{total}}{N_o} = \frac{I_{own}+I_{other}+N_o}{N_o} \qquad (8)$$

The cell load n, is defined as the ratio of the received power from all active users to the total received power as given in equation (9). The cell load threshold must not be exceeded for call admission at any given state of the cell.

$$n = \frac{I_{own}+I_{other}}{I_{own}+I_{other}+N_o} \qquad (9)$$

The noise rise is related to the cell load given in equation (10) as follows:

$$NR = \frac{I_{total}}{N_o} = \frac{I_{own}+I_{other}+N_o}{N_o} = \frac{1}{\frac{N_o}{I_{own}+I_{other}+N_o}} = \frac{1}{1-\frac{I_{other}+N_o}{I_{own}+I_{other}+N_o}} = \frac{1}{1-n} \qquad (10)$$

Thus, instead of using noise rise, the cell load can be used to determine call admission.

Stage 4: Computation of Soft Blocking Probability when Cell load Threshold is Exceeded

Simulation Model: In CDMA network, the cell load, n is interpreted as shared resource and the load per cell (loading factor), L_k as resource requirement. The application of the Kaufman-Roberts algorithm postulates a discrete shared resource and discrete service requirements. Thus, in order to calculate the new call blocking probabilities of different service classes, we make discrete the loading factor and the cell load by introducing a cell load unit g of which n_{max} is an integer multiple. This will help us determine the system state probabilities. Thus, the resulting capacity and resource requirements are:

$$C = \frac{n_{max}}{g} \quad \text{and} \quad r_k = \text{round}\left(\frac{L_k}{g}\right)$$

Since a state corresponds to the resources occupied when all users are active. We denote by $c = n_{own}/g$ as the number of occupied resources by the active users and introduce a random variable Λ for the number of occupied resources. Still assuming no local blocking occurs, the probability $\Lambda(c|i)$ that c resources are occupied in state i (bandwidth occupancy) is computed from equation (11):

$$\Lambda(c|i) = \sum_{k=1}^{K} P_k(i) \left[v_k \Lambda(c - r_k | i - r_k) + (1 - v_k)\Lambda(c|i - r_k)\right] \tag{11}$$

for $i = 1, \dots, i_{max}$ and $c \leq i$, where i_{max} is the highest reachable system state.

The resource occupancy distribution is computed according to equation (11). So, using the theorem of total probability we derive the local blocking probability (blocking factor) as in [8] as:

$$P_{b,k}(i) = \sum_{c=0}^{i} \Lambda(c|i) P_{b,k}(c) \tag{12}$$

Again, we denote by $P_k(i)$, the probability that the system is in state i or the probability that state i is reached by a new call of service k or the probability that state i is reached from state $i - r_k$ as follows:

$$P_k(i) = \frac{\mathcal{P}(i-r_k)\left(1-P_{b,k}(i-r_k)\right)\alpha_k\frac{r_k}{i}}{\sum_{i=0}^{c}\mathcal{P}(i)} \tag{13}$$

where,

$$\mathcal{P}(i) = \begin{cases} 1, & for\ i = 0 \\ \sum_{k=1}^{K} \mathcal{P}(i - r_k)\left(1 - P_{b,k}(i - r_k)\right)\alpha_k\frac{r_k}{i} & for\ i = 1, \dots, C \\ 0, & else \end{cases} \tag{14}$$

This probability depends on $\mathcal{P}(i - r_k)$ and $P_{b,k}(i - r_k)$ which are known for all states c with $c < i$. $\alpha_k, r_k, \mathcal{P}(i), P_{b,k}(i)$ are the parameters of the model with infinite number of sources. Still observing the steady state probabilities given in equation (14), we obtain the total (soft) blocking probability for a service-class k as the sum of all state probabilities $P_k(i)$ multiplied with the blocking probabilities $P_{b,k}(i)$, for all reachable states as shown in equation (15):

$$P_{Sb(k)} = \sum_{i=0}^{i_{max}} P_{b,k}(i) * P_k(i) \tag{15}$$

Stage 5: Blocking or Accepting New Call based on Computed Probability

The developed model is aimed at enhancing the acceptance of more user calls into the system to improve overall utilization of scare network resources. Thus, a call is blocked when the blocking probability value is greater than the predefined threshold of 0.01given in [15] or at worst 0.02, otherwise, it is accepted, assigned a channel with needed QoS requirement and the channel released on call completion.

In [22], an intelligent CAC scheme was developed where fuzzy logic technique was adopted in the admission decision. Calls admissions were termed strongly accepted, weakly accepted, weakly rejected and strongly rejected depending on the output value of the multi-criteria parameters in the input to the call admission controller. Their results show the capability of fuzzy logic to improve system performance by accepting more user calls.

Stage 6: Storage of Generated Results and Updated System Parameters

At this stage, the result obtained from the computed soft blocking probability which is used to determine the acceptance or blocking of the new call is stored. Similarly, system parameters generated are also stored in the database so that the next state of a cell can be determined from parameters from the previous state. The objective is to improve battery live, reduce inter-cell interference, maintain desired service quality, and maximize utilization of network resource for improved system performance.

4. MODEL IMPLEMENTATION

Figure 4 shows the pseudo code for the developed algorithm. The model was validated through simulations in MATLAB and its equations were implemented in Java programming language. The results obtained are as shown in figures 5-7 for the three performance measures listed below. Table 1 indicates the simulation parameters.

Performance Measure 1 ($P_{sb(k)}$ vs. α_k at different R_k values): The impact of data rates on soft blocking for given offered load.

Performance Measure 2 ($P_{sb(k)}$ vs. α_k at different E_b/N_o values): The impact of signal energy per bit to noise spectral density for given offered load.

Performance Measure 3 ($P_{sb(k)}$ vs. α_k at different I_{other} values): The impact of other-cell interference on soft blocking for given offered load.

In figure 5, the developed system accepts input parameters and runs numbers of simulations for performance measure 1(i.e. model 1). The result obtained indicates that at different traffic data rates for given offered load to the system, the soft blocking probability values are higher for services that require higher data rates and vice versa. Nevertheless, the overall performance of the system is improved as the soft blocking probability values obtained are less than 0.02 showing significant reduction.

In figure 6, the system evaluates the impact of each user's signal energy per bit to noise spectral density E_b/N_o on soft blocking since in reality, it value cannot be fixed for all service classes in a CDMA network. This parameter is sometimes used to ascertain the user's QoS requirement. For different values of E_b/N_o input for performance measure 2 (i.e. model 2), the soft blocking probability is reduced significantly indicating that more user calls are admitted into the system for resource sharing.

In figure 7, the impact of other-cell interference I_{other} on soft blocking probability is simulated for given traffic load considering performance measure 3 (i.e. model 3). The results indicate that though the soft blocking probability values obtained are higher for higher values of I_{other}

indicating the negative effect of inter-cell interference on CDMA networks, the values are significantly reduced thus allowing lower blocking rates.

```
01        while (connection request is from a new user) {
02        do
03              initiate new call request;
04        if (I_total ≤ I_max < I_th)
05              accept new call request;
06        elseif (n + L_k ≤ n_max)
07              accept new call request;
08        else
09              { compute P_sb(k);
10        if (P_sb(k) ≤ P_sb(k)th)
11              accept new call request;
12        else
13        Block new call request; } }
14        Ignore connection request; // it's a handoff request
15        End
```

Figure 4: Pseudo code for the developed algorithm

Table 1: Simulation parameters

Parameter	Value
Chip rate, W	5MHz
BS thermal noise, N_o	-174dBm
Data rate, R_k	144Kbps,253Kbps,384Kbps
Voice activity factor, v_k	0.65, 1
Signal bit energy over noise spectral density, E_b/N_o	3 - 7dB
Multi user detection factor, β	0.78
Transmitting antenna gain, G_{tx}	11dBi
BS antenna height, Δh_b	15m, omnidirectional
MS transmitted power signal	21-30dBm
Maximum cell load, n_{max}	0.8
Values of other-cell interference, I_{other}	$5\,e^{-19}mW$ - $6e^{-18}mW$
Offered load, α_k	0.2 – 2.0 Erlang
Loading factor, L_k	0.05, 0.1, 0.15, 0.2, 0.25
Maximum number of states, i_{max}	4
Number of service class, k	3

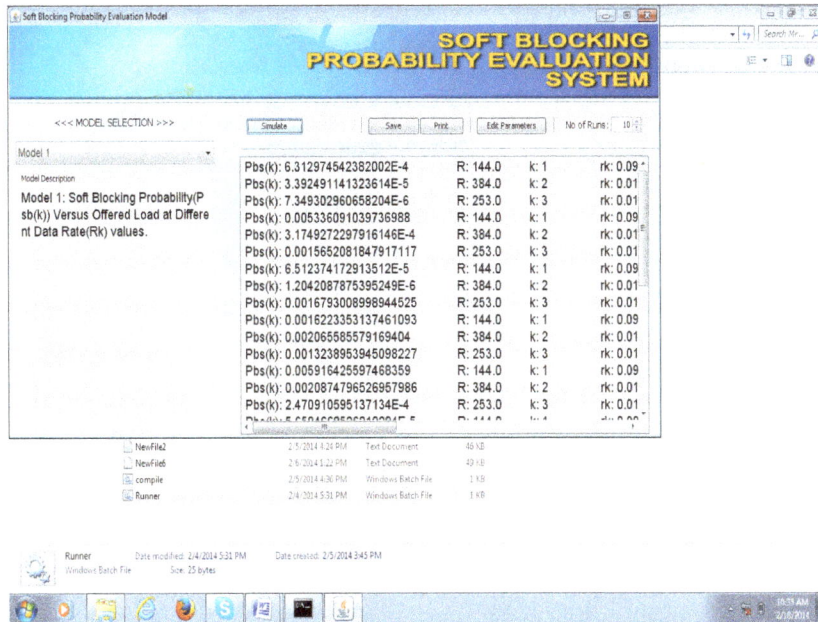

Figure 5: Soft blocking results for performance measure 1

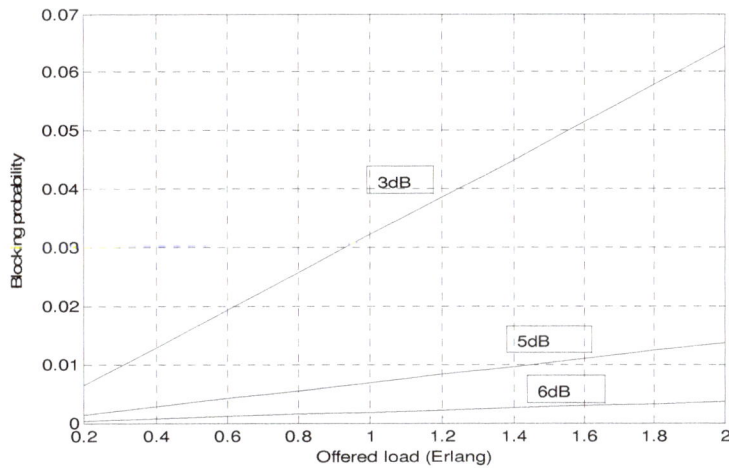

Figure 6a: Soft blocking results for performance measure 2 in MATLAB

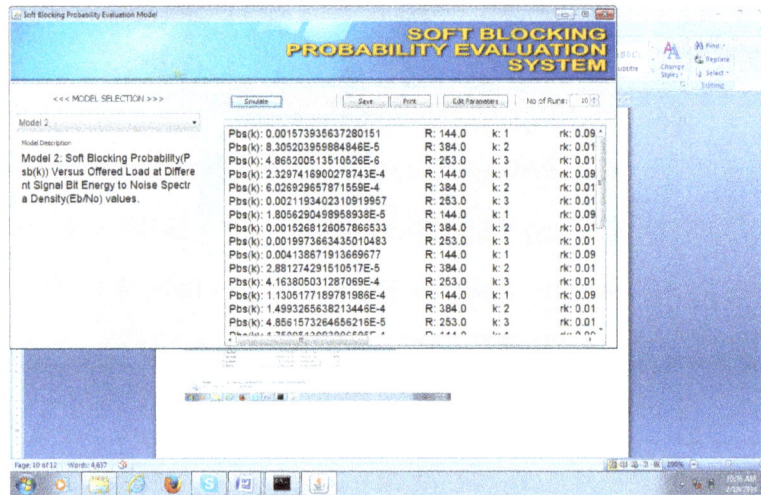

Figure 6b: Soft blocking results for performance measure 2 in JAVA

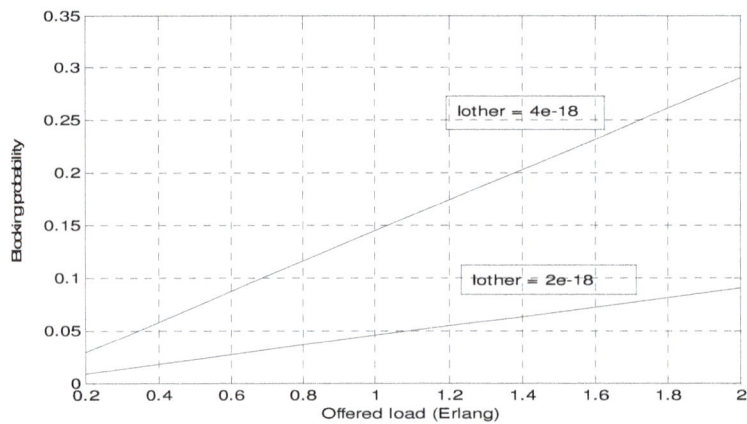

Figure 7a: Soft blocking results for performance measure 3 in MATLAB

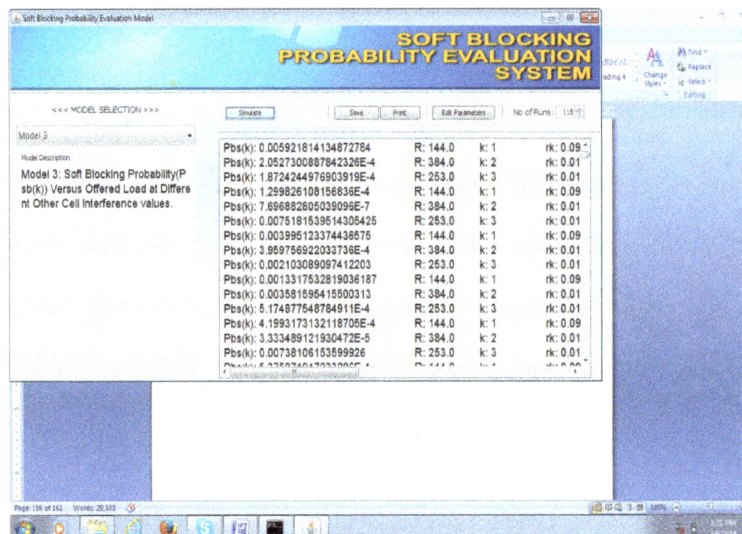

Figure 7b: Soft blocking results for performance measure 3 in JAVA

5. CONCLUSION

We started from the verification that call blocking perceived negatively by users in mobile cellular networks actually exist at an alarming rate in a well established CDMA network - Visafone in the region under study. As a solution, we developed a model for soft blocking probability evaluation by extending the Kaufman-Roberts algorithm which allows an efficient approximation of the blocking probabilities. We validate our results by simulation and show that the approximation yields accurate results even for large other-cell interferences and low user activities. Moreover, the impact of the model's parameters on soft blocking probability has been studied. This work showed that an efficient and fair resource management is possible for supporting traffic with strict QoS requirements. The results are particularly useful for operators who aim to predict the QoS of their networks for several combinations of the parameters (for dimensioning, prediction or optimization).

The imperfect power control framework considered addresses three main issues while ensuring that MSs achieve their QoS targets. First, it reduces the overall interference from neighboring cells, which is important for frequency reuse in multi-cell CDMA network. Second, it reduces the battery usage of MSs according to their individual preferences. Third, it mitigates the near-far problem by ensuring that MSs closest to the BS do not overpower the system at the detriment of those farther away.

REFERENCES

[1] Ramesh Babu H.S., Gowrishankar & Satyanarayana P.S. (2009) "Call Admission Control Performance Model for Beyond 3G Wireless Networks", (IJCSIS) International Journal of

Computer Science and Information Security, Vol. 6, No. 3, pp224-229.

[2] Karray, M. K. (2010) "Evaluation of the Blocking Probability and the Throughput in the Uplink of Wireless Cellular Networks", IEEE, In Proceedings of International Conference on

Communications and Networking (CommNet), Tozeur, Tunisia.

[3] Levine, D., Akyildiz, I. & Naghshineh, M. (1997) "A Resource Estimation and Call Admission Algorithm for Wireless Multimedia Networks using the Shadow Cluster Concept", IEEE/ACM

Trans. Net., Vol. 5, No. 1, pp1–12.

[4] Epstein B. & Schwartz, M. (2000) "Predictive QoS-based Admission Control for Multiclass Traffic in Cellular Wireless Networks", IEEE JSAC, Vol. 18, No. 3, pp523–534.

[5] Choi, S. & Shin, K. G. (2002) "Adaptive Bandwidth Reservation and Admission Control in QoS-Sensitive Cellular Networks". IEEE Transactions on Parallel and Distributed Systems, Vol. 13.

[6] Chang, J., Chung, J. & Sung, D. (2006) "Admission Control Schemes for Soft Handoff in DS-CDMA Cellular Systems Supporting Voice and Stream-type Data Services", IEEE Trans.Vehic. Tech., Vol. 51, No.6, pp1445–14459.

[7] Yang, X. & Bigham, J. (2007) "A Call Admission Control Scheme using NeuroEvolution Algorithm in Cellular Networks", IJCAI, pp186-191.

[8] Vassilakis, V. G. & Logothetis, M. D. (2008) "The Wireless Engset Multi-Rate Loss Model for the Handoff Traffic Analysis in WCDMA Networks", IEEE.

[9] Parkinson, R. (2005) Traffic Engineering Techniques in Telecommunications, Infotel Systems Inc. Retrieved 2012-10-17 from http://www.tarrani.net/mike/docs/TrafficEngineering.pdf

[10] Zukerman, M. (2008) An Introduction to Queuing Theory and Stochastic Teletraffic Models Retrieved 2012-11-27 from http://www.ee.cityu.edu.hk/~zukerman/classnotes.pdf

[11] Quo, J., Chan, S., Wong, E. W. M., Zukerman, M., Taylor, P. & Tang, K. S. (2003) "On Blocking Probability Evaluation for Video-on-Demand Systems".

[12] Kaufman, J. (1981) "Blocking in a Shared Resource Environment", IEEE Trans.Commun., Vol. 29, No. 10, pp1474-1481.

[13] Roberts, J. W. (1981) "A Service System with Heterogeneous User Environments", in G. Pujolle (Ed.), Performance of Data Communications Systems and their Applications, North-Holland, Amsterdam, pp423-431.

[14] Andrew, L. L. H., Payne, D. J. B. & Hanly, S. V. (1999) "Queuing Model for Soft-blocking CDMA Systems", IEEE, pp436-440.

[15] Verdone, R., Orriss, J., Zanella, A. & Barton, S. K. (2002) "Evaluation of the Blocking Probability in a Cellular Environment with Hard Capacity: A Statistical Approach", In Proceedings of 13th IEEE International Symposium on person, Indoor and Mobile Radio Communications, Vol. 2: pp658-622.

[16] Bonald, T. & Proutiere, A. (2005) "Conservative Estimates of Blocking and Outage Probabilities in CDMA Networks". Performance Evaluation, Vol. 62, No. 14, pp50-67.

[17] Razavilar, J. Farrokhi, F. R. & Liu, K. J. R., (2002) "Blocking Probability of Handoff Calls and Carried Traffic in Wireless Networks with Antenna Arrays", In Proceedings of 1st Asihomer Conference on Signals, Systems and Computers, Vol. 1, pp635-639.

[18] Huang, J., Huang, C. Y. & Chou, C. M. (2004) "Soft-blocking Based Resource Dimensioning for CDMA Systems", IEEE Veh. Tech. Conf. VTC2004, Vol.6, pp4306-4309.

[19] Kallos, G. A., Vassilakis, V. G. & Logothetis, M. D. (2008) "Call blocking probabilities in a W-CDMA Cell with Fixed Number of Channels and Finite Number of Traffic Sources".

[20] Hou, I. & Kumar, P. (2009) "Admission Control and Scheduling for QoS Guarantees for Variable Bit-Rate Applications on Wireless Channels", in Proc. of MobiHoc'09, pp175-184.

[21] 3GPP TR 25.942 v10.0.0 (2011). Radio Frequency (RF) System Scenarios.

[22] Asuquo, D. E., Williams, E. E., Nwachukwu, E. O. & Inyang, U. G. (2013) "An Intelligent Call Admission Control Scheme for Quality of Service Provisioning in a Multi-traffic CDMA Network", International Journal of Scientific and Engineering Research, Vol. 4, No. 12, pp152-161.

Permissions

The contributors of this book come from diverse backgrounds, making this book a truly international effort. This book will bring forth new frontiers with its revolutionizing research information and detailed analysis of the nascent developments around the world.

We would like to thank all the contributing authors for lending their expertise to make the book truly unique. They have played a crucial role in the development of this book. Without their invaluable contributions this book wouldn't have been possible. They have made vital efforts to compile up to date information on the varied aspects of this subject to make this book a valuable addition to the collection of many professionals and students.

This book was conceptualized with the vision of imparting up-to-date information and advanced data in this field. To ensure the same, a matchless editorial board was set up. Every individual on the board went through rigorous rounds of assessment to prove their worth. After which they invested a large part of their time researching and compiling the most relevant data for our readers.

The editorial board has been involved in producing this book since its inception. They have spent rigorous hours researching and exploring the diverse topics which have resulted in the successful publishing of this book. They have passed on their knowledge of decades through this book. To expedite this challenging task, the publisher supported the team at every step. A small team of assistant editors was also appointed to further simplify the editing procedure and attain best results for the readers.

Apart from the editorial board, the designing team has also invested a significant amount of their time in understanding the subject and creating the most relevant covers. They scrutinized every image to scout for the most suitable representation of the subject and create an appropriate cover for the book.

The publishing team has been an ardent support to the editorial, designing and production team. Their endless efforts to recruit the best for this project, has resulted in the accomplishment of this book. They are a veteran in the field of academics and their pool of knowledge is as vast as their experience in printing. Their expertise and guidance has proved useful at every step. Their uncompromising quality standards have made this book an exceptional effort. Their encouragement from time to time has been an inspiration for everyone.

The publisher and the editorial board hope that this book will prove to be a valuable piece of knowledge for researchers, students, practitioners and scholars across the globe.

List of Contributors

Syed Shakeel Hashmi
Electronics and Communication Engineering, FST ICFAI University Dehradun, India

Yun Wang
School of Computer Science and Engineering, CNII, Southeast University, Nanjing, 211189, P.R. China

Peizhong Shi
School of Computer Science and Engineering, CNII, Southeast University, Nanjing, 211189, P.R. China

Kai Li
School of Computer Science and Engineering, CNII, Southeast University, Nanjing, 211189, P.R. China

Jie Wu
Department of Computer and Information Sciences, Temple University, Philadelphia, PA 19122

Mohammad A Hoque
Department of Computer Science, The University of Alabama, USA

Xiaoyan Hong
Department of Computer Science, The University of Alabama, USA

Pavani Sanghoi
Student, Department of Electronics and Communication Engineering, Lovely Professional University, Punjab, India

Lavish Kansal
Assistant Professor, Department of Electronics and Communication Engineering, Lovely Professional University, Punjab, India

Thomas Bourgeois
Graduate School of Global Information and Telecommunication Studies Waseda University, Japan

Shigeru Shimamoto
Graduate School of Global Information and Telecommunication Studies Waseda University, Japan

S. F. Yunas
Department of Electronics and Communications Engineering, Tampere University of Technology, Tampere, Finland

T. Isotalo, J. Niemelä
Department of Electronics and Communications Engineering, Tampere University of Technology, Tampere, Finland

M. Valkama
Department of Electronics and Communications Engineering, Tampere University of Technology, Tampere, Finland

Arindam Sarkar
Department of Computer Science & Engineering, University of Kalyani, W.B, India

J. K. Mandal
Department of Computer Science & Engineering, University of Kalyani, W.B, India

Mustafa Abu Nasr
Engineering Department, Al Azhar University, Gaza, Palestine

Mohamed K. Ouda
Electrical Engineering Department, Islamic University of Gaza, Gaza, Palestine

Samer O. Ouda
Electrical Engineering Department, Islamic University of Gaza, Gaza, Palestine

Khaled Almi'ani
Al-Hussein Bin Talal University, Ma'an, Jordan

Muder Almi'ani
Al-Hussein Bin Talal University, Ma'an, Jordan

Ali Al_ghonmein
Al-Hussein Bin Talal University, Ma'an, Jordan

Khaldun Al-Moghrabi
Al-Hussein Bin Talal University, Ma'an, Jordan

Hakan Koyuncu
Computer Science Department, Loughborough University, Loughborough, UK

Shuang Hua Yang
Computer Science Department, Loughborough University, Loughborough, UK

Neelesh Gupta
Research Scholar, UIT, RGPV, Bhopal (M.P.)-India

Roopam Gupta
Dept. of IT, UIT, RGPV, Bhopal (M.P.)-India

Md. Humayun Kabir
Department of Computer Science & Engineering, University of Rajshahi, Bangladesh

Edem E. Williams
Department of Mathematics/Statistics and Computer Science University of Calabar, PMB 1115, Calabar-Nigeria

Daniel E. Asuquo
Department of Computer Science University of Uyo, PMB 1017, Uyo-Nigeria